Industrial Research and Technological Innovation

An Econometric Analysis

Industrial Research and Technological Innovation

An Econometric Analysis

EDWIN MANSFIELD
PROFESSOR OF ECONOMICS
WHARTON SCHOOL
UNIVERSITY OF PENNSYLVANIA

Published for the Cowles Foundation for Research in Economics at Yale University

W. W. NORTON & COMPANY, INC.
NEW YORK

Library of Congress Catalog Card No. 67-11086

Published simultaneously in Canada by
George J. McLeod Limited, Toronto

PRINTED IN THE UNITED STATES OF AMERICA

2 3 4 5 6 7 8 9 0

To L. H. M.

E. D. M.

E. D. M.

Contents

Figures

Tables

Preface

After many years of neglect, technological change is receiving the attention from economists that it deserves, the 1960's being a time of enormous interest in this area in academic, government, and business circles. Central to the economics of technological change is the manner in which new processes and products are conceived, developed, commercialized, and accepted. To help promote a better understanding of this process, I have been engaged for a number of years in a series of related econometric studies of industrial research and technological innovation. The purpose of this book is to bring together the results of these studies.

This book is addressed primarily to a professional audience of economists, engineers, and scientists, as well as to relatively advanced students in these areas; other students and laymen may find my *The Economics of Technological Change* more suitable to their background and tastes. A companion volume to this one, the latter book is an overview and interpretation of this entire field, including a discussion and analysis of public and private policy issues. It covers a wider spectrum of topics, and is technically less demanding than this book, which is devoted entirely to a very detailed and intensive investigation of a relatively small number of important questions.

Acknowledgments

Without the very generous financial support of the National Science Foundation, these studies could not have been carried out. I have also received very substantial support from the Ford Foundation, as a member of the six-man Inter-University Committee on Technological Change and Economic Growth and as a recipient during 1960–1961 of a Ford Foundation Faculty Research Fellowship. In addition, I appreciate very much the financial support and encouragement of the Cowles Foundation. I spent parts of 1961 and 1962 at the Foundation, and this book is a Cowles publication. These studies were begun while I was a member of the faculty at the Carnegie Institute of Technology's Graduate School of Industrial Administration, they continued while I was at Yale and Harvard Universities, and they were completed at the University of Pennsylvania's Wharton School. I am very much indebted to my colleagues and students at Carnegie, Yale, Harvard, and Wharton for their advice and assistance.

The book has also benefited greatly from experience I gained while acting as a consultant in this area to The RAND Corporation; the President's Office of Science and Technology; the National Commission on Technology, Automation, and Economic Progress; the U.S. Army Ballistics Research Laboratories; the Assistant Secretary of Commerce for Science and Technology; the White House Panel on Civilian Technology; the U.S. Interagency Energy Study; the Ohio Research and Development Foundation; the University of Maryland; the Denver Research Institute; the Federal Power Commission; and a number of industrial firms. The book also benefited from insights gained while serving on the Governor's Science Advisory Committee.

When it comes to acknowledging the help of particular individuals, the task is extremely difficult, since hundreds of people have contributed in various ways to the completion of this book. Without the generous cooperation of the many executives who provided data and advice, without the help of the many economists, statisticians, and operations researchers who commented on the results, without the encouragement of a number of administrators, the book could not have been finished. Nonetheless, there are a few people to whom I owe a particularly great debt. First, there is Lee Bach, who encouraged me to work in this area. Second, there are Herbert Simon, Tjalling Koopmans, John Dunlop, and Burton Klein, all of whom provided encouragement and advice. Third, there are Zvi Griliches, Richard Nelson, and Karl Shell, all of whom read the manuscript and provided useful comments. Most important, there is my wife, who contributed in countless ways to the completion of the work.

Finally, I want to thank the following journals for permission to include material that first appeared in their pages: *American Economic Review*, *Journal of Political Economy*, *Quarterly Journal of Economics*, *Econometrica*, *Journal of Business*, and *Review of Economics and Statistics*. Preliminary versions of various chapters were published in these journals to stimulate discussion and constructive criticism. A list of these articles can be found in the References; their titles correspond closely to the titles of the chapters. Of course, some of this material has been revised and new material has been added.

Part I

Introduction

CHAPTER 1

Basic Concepts and Facts

There is general agreement among economists and others that one of the most powerful forces influencing the American economy is technological change—the advance in knowledge relative to the industrial arts which permits, and is often embodied in, new methods of production, new designs for existing products, and entirely new products and services. The National Commission on Technology, Automation, and Economic Progress goes so far as to state: "It is easy to oversimplify the course of history; yet if there is one predominant factor underlying current social change, it is surely the advancement of technology." [1] Of course, this does not mean that all effects of technological change have been beneficial. Although technological change has improved working conditions and added many new dimensions to life, it has also made possible the destruction of mankind on an unprecedented scale, thrown whole communities into distress, and polluted certain parts of the environment. Nonetheless, most economists would agree that technological change has, on balance, been beneficial—and all would agree that it is an important area to study.

In recent years, economists—as well as other social scientists, physical scientists, government officials, businessmen, labor leaders, and others—have shown a very great interest in technological change. Indeed this has become one of the most fashionable areas of economics, due largely to the growing awareness that our rate of economic growth depends heavily on our rate of technological change; to the obvious dependence of our national security on the output of our military research and development effort; to the growing realization of the importance of competition through new products and processes rather than direct price competition; to the claims made by some observers that "automation" will lead to widespread unemployment and the need for large retraining programs; and to the concern in many quarters regarding the adequacy of our national policies toward science and technology.

[1] See *Technology and the American Economy*, Report of the National Commission on Technology, Automation, and Economic Progress (Washington, 1966), p. xi.

1. Technological Change and the Production Function

The purpose of this chapter is to present some basic material regarding the economics of technological change.[2] To understand the econometric studies described in subsequent chapters, one must be familiar with this material. We begin by defining technological change in a particular industry. At any point in time there exist a number of ways of producing the industry's product, some using little capital and much labor, some using much capital and little labor, some cheap, some expensive, some old, some new. Each of these alternative methods of producing the product is called a production possibility. The production function shows the maximum output rate which can be obtained from given amounts of inputs (like labor and capital). To obtain the production function, one omits those production possibilities that are technologically inefficient—in the sense that to produce the given quantity of output they use more of one input, and at least as much of other inputs, than some other process.

Technological change results in a change in the production function. If the production function were readily observable, a comparison of its position at two points in time would provide the economist with a simple measure of the effect of technological change during the intervening period. Of course, the change in the production function may depend on improvements of various sorts—a better piece of equipment, an improved material, better principles of organization, and so on. Technological change also results in the availability of new products. In many cases, the availability of new products can be regarded as a change in the production function since they are merely more efficient ways of meeting old wants, if these wants are defined with proper breadth. In other cases, however, the availability of new products cannot realistically be viewed as a change in the production function since they entail an important difference in kind.

It is important to distinguish between technological change and change in techniques. A technique is a utilized method of production. Thus, whereas a technological change is an advance in knowledge, a change in technique is an alteration of the character of the equipment, products, and organization which are actually being used. Technological change should also be distinguished from scientific advance. Technological change often occurs as a result of inventions that do not depend on new scientific principles. Until the middle of the nineteenth century, when research methods were first used in a systematic way to develop new products in the field of chemistry, little practical use was made of scientific principles. The inventions that provided the basis for the industrial revolution were invented by practical men and based upon observation, art, and common sense. Turning to the present, it

[2] For a much more complete discussion of the topics taken up in this and succeeding sections of this chapter, see E. Mansfield, *The Economics of Technological Change* (New York: Norton, 1968).

is still true that many changes in technology require no new scientific principles.

2. Measurement of Technological Change

Perhaps the simplest and most commonly used measure of the rate of technological change is the change in output per manhour. It is a partial productivity index, in the sense that output is related to one input: labor, without recognition of changes in the quantity of other inputs. Unfortunately, it is a very incomplete measure of the rate of technological change since it reflects a great many factors other than technological change. For example, it is affected by increases in the amount of capital per worker, economies of scale, changes in the extent to which productive capacity is used, and the rate of diffusion of best-practice techniques.

Another measure that is commonly used is the so-called total productivity index, which relates changes in output to changes in both labor and capital inputs, not changes in labor inputs alone. Specifically, this index equals

$$\frac{q}{zl + vk},$$

where q is output (as a percent of output in some base period), l is labor input (as a percent of labor input in some base period), k is capital input (as a percent of capital input in some base period), z is labor's share of the value of output in the base period, and v is capital's share of the value of the output in the base period. This index has several advantages over output per manhour, the most important being that it takes account of change over time in the amount of capital inputs. It is less likely to be influenced by capital-labor substitution due to changes in factor prices. However, it has the disadvantage of assuming that the marginal products of the inputs are altered only by technological change and that their ratios remain constant and independent of the ratios of the quantities of the inputs.[3]

Still other measures have been developed as well. Solow, in an important paper [154] published in 1957, provided a measure based on the assumption that there are constant returns to scale, that capital and labor are paid their marginal products, and that technological change is neutral. In 1959, he [155] suggested a somewhat different measure based on the assumption that all technological change must be embodied in new capital if it is to be utilized; that is, all technological change is capital-embodied. More recently, Arrow, Chenery, Minhas, and Solow [6] provided a measure of the rate of technological change based on a constant-elasticity-of-substitution production function.

[3] See E. Domar, "On Total Productivity and All That," *Journal of Political Economy* (December, 1962). For an alternative definition of the total productivity index, see his equation (3).

What do these measures suggest? As an example, consider the total productivity index. Results based on this index indicate that during 1889–1957, the rate of productivity growth in the private economy averaged about 1.7 percent per year. However, it seems to have been higher after World War I than before, and higher during periods of expansion than during periods of contraction. When international differences are considered, the rate of productivity growth seems to have been higher after World War II in Japan and Germany than in the United States and Canada, and higher in the United States and Canada than in the United Kingdom. Within the United States, it seems to have been higher during 1899–1957 in transportation and communication than in mining, manufacturing, and farming. Within U.S. manufacturing it seems to have been highest in rubber, transportation equipment, tobacco, chemicals, printing, glass, fabricated metals, textiles, and petroleum [66].[4]

Even the most sophisticated of the available measures of the rate of technological change suffers from very important limitations. Because technological change is measured by its effects, and its effects are measured by the growth of output unexplained by other factors, it is impossible to sort out technological change from the effects of whatever inputs are not included explicitly in the analysis. The customary measures are plagued by the difficult problems, both theoretical and practical, in evaluating entirely new products. Moreover, it is sometimes impossible to distinguish capital-embodied from disembodied technological change. In addition, when one compares various studies, there is considerable variation in the estimated rates of technological change in particular industries. Because of these and other problems, the available measures should be used only as very rough guides.

3. Determinants of the Rate of Technological Change

What determines the rate of technological change in an industry? On a priori grounds, one would expect it to depend to a large extent on the amount of resources devoted by firms, independent inventors, and government to the improvement of the industry's technology. The amount of resources devoted by the government depends on how closely the industry is related to the defense, health, and other social needs for which the government assumes major responsibility, on the extent of the external economies generated by the relevant research and development, and on more purely political factors. The amount of resources devoted by industry and independent inventors depends heavily on the profitability of their use. Econometric studies (described in Chapters 2 and 3) indicate that the total amount a firm spends on research and development is influenced by the expected profitability of the R and D projects under consideration, and that the probability of its accepting a particular R and D project depends on

[4] Of course, results based on other measures (rather than the total productivity index) differ from those presented above. See Mansfield, *Economics of Technological Change*, Chapter 2.

the project's expected returns. Case studies of particular inventions and studies of patent statistics seem to support this view.[5]

Accepting the proposition that the amount invested by private sources in improving an industry's technology is influenced by the anticipated profitability of the investment, it follows that the rate of technological change in a particular area is influenced by the same kinds of factors that determine the output of any good or service.[6] On the one hand, there are demand factors which influence the rewards from particular kinds of technological change. For example, if a prospective change in technology reduces the cost of a particular product, increases in the demand for the product are likely to increase the returns from effecting this technological change. Similarly, a growing shortage and a rising price of the inputs saved by the technological change are likely to increase the returns from effecting it. As an illustration, consider the history of English textile inventions. During the early eighteenth century, there was an increase in the demand for yarn, due to decreases in the price of cloth and increased cloth output. This increase in demand, as well as shortages of spinners and increases in their wages, raised the returns to inventions that increased productivity in the spinning processes and directly stimulated the work leading to such major inventions as the water frame, the spinning jenny, and the spinning mule.

On the other hand, there are also supply factors which influence the cost of making particular kinds of technological change. Obviously, whether people try to solve a given problem depends on whether they think it can be solved and how costly it will be, as well as on the payoff if they are successful. The cost of making science-based technological changes depends on the number of scientists and engineers in relevant fields and on advances in basic science; for example, advances in physics reduced the cost of effecting changes in technology in the field of atomic energy. In addition, the rate of technological change depends on the amount of effort devoted to making modest improvements that lean heavily on practical experience. Although there is often a tendency to focus attention on the major, spectacular inventions, it is by no means certain that technological change in many industries is due chiefly to these inventions, rather than to a succession of minor improvements; for example, Gilfillan has shown that technological change in shipbuilding has been largely the result of gradual evolution. In industries

[5] For example, see M. Peck, "Inventions in the Postwar American Aluminum Industry," T. Marschak, "Strategy and Organization in a System Development Project," and R. Nelson, "The Link Between Science and Invention: The Case of the Transistor," all in *The Rate and Direction of Inventive Activity* (Princeton: Princeton University, 1962).

[6] Needless to say, these factors are not the only ones that influence the rate of technological change. As emphasized below and in subsequent chapters, there is considerable uncertainty in the research and inventive processes, and laboratories, scientists, and inventors are motivated by many factors other than profit. Nonetheless, the "economic" factors seem important. For further discussion, see J. Schmookler, *Invention and Economic Growth* (Cambridge: Harvard University Press, 1966) and R. Nelson, M. Peck, and E. Kalachek, *Technology, Economic Growth, and Public Policy* (Washington: Brookings Institution, 1967).

where this is a dominant source of technological change and where technological change is only loosely connected with scientific advance, one would expect the rate of technological change to depend on the number of people working in the industry and in a position to make an improvement of this sort.[7]

Besides being influenced by the quantity of resources an industry devotes to improving its own technology, an industry's rate of technological change depends on the quantity of resources devoted by other industries to the improvement of the capital goods and other inputs it uses. Technological change in an industry that supplies components, materials, and machinery often prompts technological change among its customers. For example, consider the case of aluminum. For about 30 years after the development of processes to separate aluminum from the ore, aluminum technology remained dormant because of the lack of low-cost electrical power. Technological change in electric power generation, due to Thomas Edison and others, was an important stimulus to the commercial production of aluminum and to further technological change in the aluminum industry. In addition, there is another kind of interdependence among industries. Considerable "spillover" occurs, techniques invented for one industry turning out to be useful for others as well. For example, continuous casting was introduced successfully in the aluminum industry before it was adapted for use in the steel industry. The inventor turned his attention to steel after inventing a process for casting nonferrous metals, which are easier to cast because of their lower melting points.

Other factors which influence an industry's rate of technological change are the industry's market structure; the legal arrangements under which it operates; the attitudes toward technological change of management, workers, and the public; the way in which the firms in the industry organize and manage their research and development; the way in which the scientific and technological activities of relevant government agencies are organized and managed; and the amount and character of the research and development carried out in the universities and in other countries. With regard to market structure, there has been considerable argument over the role of the very large firm. Some, like Galbraith [40] and Schumpeter [149], argue that such firms are needed to produce the technical achievements on which economic progress depends. Others argue the contrary. More will be said on this score in subsequent chapters.

4. Expenditures on Research and Development: Growth and Flow of Funds

"Research" is original investigation directed to the discovery of new scientific knowledge, and "development" is technical activity concerned with

[7] See S. Gilfillan, *Inventing the Ship* (Follett, 1935); S. Hollander, *The Sources of Increased Efficiency* (Cambridge: M.I.T. Press, 1965); and K. Arrow, "The Economic Implications of Learning by Doing," *Review of Economic Studies* (June 1962).

nonroutine problems encountered in translating research findings into products and processes. Although there is no clear line between research and development, they are by no means the same thing. Whereas research is conducted to obtain new knowledge, development is required to reduce the knowledge to practice. This often entails the making of various types of experiments, the design and development of prototypes, and the construction of pilot plants.

Recent years have witnessed a tremendous growth in the amount spent on research and development. In 1945, industry performed about $1.2 billion worth and by 1963 about $12.7 billion; in 1945, about $400 million worth

Table 1.1. *Total R and D Expenditures and Number of Research Scientists and Engineers, U.S., 1941–1963*

YEAR	TOTAL R AND D EXPENDITURES ($ millions)	NUMBER OF RESEARCH SCIENTISTS AND ENGINEERS (in thousands)
1941	900	87
1943	1,210	97
1945	1,520	119
1947	2,260	125
1949	2,610	144
1951	3,360	158
1953	5,160	223[b]
1955	6,200	n.a.
1957	9,810	327[c]
1959	12,430	n.a.
1961	14,380	387
1963[a]	17,350	n.a.

SOURCE: *The Growth of Scientific Research and Development*, Department of Defense, 1953, pp. 10 and 12; *National Science Foundation Review of Data on Research and Development*, No. 33, Apr. 1962; and *National Science Foundation Review of Data on Science Resources*, Vol. I, No. 4, May 1965.

[a] Preliminary.

[b] 1954 Figure.

[c] 1958 Figure.

n.a. Not available.

was performed by government and by 1963, about $2.4 billion; and in 1945, the universities and other nonprofit organizations performed about $200 million worth and by 1963, about $2.2 billion. Table 1.1 shows the enormous increase in total research and development expenditures in the United States during the 1940's and 1950's. The rate of growth is truly phenomenal.[8]

Paralleling this increase in research and development expenditures, there has been a great increase in the number of engineers and scientists engaged in this work. In 1941, there were less than 90,000; in 1961, almost 400,000 (Table 1.1). Although the number of engineers and scientists engaged in research and development has increased at an impressive rate, it has not increased as rapidly as have the expenditures in this area. This is because the increases in demand for research personnel have resulted in higher salaries, and because less skilled labor and equipment seem to have been substituted where possible for the time of engineers and scientists.

Table 1.2. *Sources of R and D Funds and Performers of R and D, by Sector, U.S., 1953 and 1963*

	R AND D PERFORMANCE, BY SECTOR				
SOURCES OF R AND D FUNDS (SECTOR)	FEDERAL GOVERNMENT	INDUSTRY	COLLEGES AND UNIVERSITIES	OTHER NONPROFIT ORGANIZATIONS	TOTAL
1953 Transfer of funds ($ millions)					
Federal government	1,010	1,430	260	60	2,760
Industry		2,200	20	20	2,240
Colleges and universities			120	—	120
Other nonprofit organizations			20	20	40
Total	1,010	3,630	420	100	5,160
1963 Transfer of funds[a] ($ millions)					
Federal government	2,400	7,340	1,300	300	11,340
Industry		5,380	65	120	5,565
Colleges and universities			260	—	260
Other nonprofit organizations			75	110	185
Total	2,400	12,720	1,700	530	17,350

SOURCE: *National Science Foundation Review of Data on Science Resources*, Vol. I, No. 4, May 1965.

[a] Preliminary figures.

Another very important thing to note is that much of the research and development *performed* by one sector is *financed* by another. Table 1.2 shows

[8] The basic source of data regarding expenditures on research and development is the National Science Foundation. See its *Reviews of Data on Research and Development*, as well as its longer reports. Of course, part of the increase in research and development expenditures is due to inflation and to shifting definitions.

that a large and increasing percentage of the research and development performed in the industrial sector is financed by the federal government. In 1953, about 40 percent was financed in this way; in 1963, about 60 percent. The situation is similar in the university sector. In 1953, about 60 percent of the research and development performed by the universities was financed by the government; in 1963, about 75 percent. In addition, the federal government financed a large and relatively stable (about 60 percent) portion of the research and development carried out by nonprofit organizations other than universities.

Besides this massive outflow of funds from the federal government to support research and development performed in other sectors, there were other flows of funds that are noteworthy, although much smaller: In recent years, industry financed about 4 percent of that carried out by universities and about 20 percent of that carried out by nonprofit organizations other than universities. Nonprofit organizations, such as the private foundations, financed about 4 percent of the research and development performed by colleges and universities.

5. Research and Development Expenditures by the Federal Government

In view of the extremely important role played by the federal government, we must look at the amount spent by various federal departments and agencies on research and development. In 1966, almost half of all federal research and development expenditures were made by the Department of Defense (Table 1.3), the primary purpose being to provide new and improved weapons and techniques to promote the effectiveness of the armed forces. The largest expenditures were made by the Air Force; the smallest were made by the Army. Only about 25 percent was spent on research (rather than development), and this research was mainly in the physical and engineering sciences. The second and third largest spenders on research and development in 1966 were the National Aeronautics and Space Administration and the Atomic Energy Commission. Both of these agencies are intimately connected with the cold war. Together with the Defense Department, they accounted for almost 90 percent of the research and development expenditures of the federal government. About one-quarter of the research and development carried out by NASA and AEC was research, most of it in the physical and engineering sciences.

In contrast with the Big Three, the fourth, fifth, and sixth largest spenders were not concerned primarily with national defense and the space race. The bulk of the research and development expenditures of the Department of Health, Education and Welfare—the fourth largest spender—was related to the work of the National Institutes of Health, the research arm of the Public Health Service. Most of the research expenditures of HEW were in the medical sciences, and about one-fifth were conducted intramurally in 1966. The fifth largest spender was the National Science Foundation, the

Table 1.3. *Federal Expenditures for R and D and R and D Plant, by Agency, Fiscal Years 1940–1966 (millions of dollars)*

DEPARTMENT OR AGENCY	1940	1948	1956	1964	1966[a]
Agriculture	29.1	42.4	87.7	183.4	257.7
Commerce	3.3	8.2	20.4	84.5	93.0
Defense	26.4	592.2	2,639.0	7,517.0	6,880.7
Army[b]	3.8	116.4	702.4	1,413.6	1,452.1
Navy[b]	13.9	287.5	635.8	1,724.2	1,540.0
Air Force[b]	8.7	188.3	1,278.9	3,951.1	3,384.4
Defense agencies	—	—	—	406.9	464.5
Department-wide funds	—	—	21.9	21.1	39.7
Health, Education and Welfare[c]	2.8	22.8	86.2	793.4	963.9
Interior	7.9	31.4	35.7	102.0	138.7
Atomic Energy Commission	—	107.5	474.0	1,505.0	1,559.7
Federal Aviation Agency	—	—	—	74.0	73.4
National Aeronautics and Space Administration[d]	2.2	37.5	71.1	4,171.0	5,100.0
National Science Foundation	—	—	15.4	189.8	258.7
Office of Scientific Research and Development	—	0.9	—	—	—
Veterans Administration	—	—	6.1	34.1	45.9
All other agencies	2.4	11.8	10.4	39.7	66.1

SOURCE: *Federal Funds for Science XIV* (National Science Foundation, 1965), Table C-46. These figures are not entirely comparable with those in previous and subsequent tables.

[a] Estimates based on requests in *The Budget, 1966.*

[b] Includes pay and allowances of military personnel support from procurement appropriations beginning in 1954.

[c] Public Health Service and Federal Security Agency prior to 1952.

[d] National Advisory Committee on Aeronautics prior to 1958.

general purposes of which are the encouragement and support of basic research and education in the sciences. Most of the Foundation's expenditures went for research in the physical and biological sciences. The sixth largest spender was the Department of Agriculture, where most of the research and development effort, which is coordinated with the research and educational activities of the land–grant colleges, was concerned with the production, utilization, and marketing of farm and forest products.

Table 1.3 shows that the total amount of research and development financed by the federal government in 1966 was over 200 times what it was in 1940 and over 4 times what it was in 1956. Future increases are likely to be much more modest. The spectacular increase in the past was due largely to wartime and postwar defense needs. The Defense Department's expenditures on research and development rose greatly during World War II and continued to increase subsequently. In the early postwar period, the Atomic Energy Commission was created and its research and development expenditures grew rapidly. During the late Fifties and early Sixties, NASA's budget grew enormously. The result has, of course, been a tremendous emphasis in the government research and development budget on defense and space technology.[9]

6. Research and Development Expenditures by Industry

In private industry, which industries are the largest performers of research and development? In recent years, research and development performance

Table 1.4. *Performance of Industrial R and D, by Industry, 1927–1961*

R AND D PERFORMANCE	1927	1937	1951	1957[a]	1961[a]
	Percent of Sales				
Aircraft and parts	n.a.	n.a.	11.9	18.9	24.2
Instruments	n.a.	n.a.	3.0	7.3	7.3
Electrical equipment	0.54	1.5	3.6	11.0	10.4
Chemicals	0.42	1.1	1.5	3.5	4.6
Rubber	0.36	0.96	0.5	n.a.	2.2
Machinery	0.19	0.43	0.5	4.2	4.4
Stone, clay and glass	0.13	0.43	0.4	n.a.	1.8
Motor vehicles	0.07	0.4	0.5	2.9 (Motor vehicles and other transportation equipment combined)	2.9 (Motor vehicles and other transportation equipment combined)
Other transportation equipment	0.07	0.07	0.3		
Primary metals and products	0.07	0.17	n.a.	n.a.	n.a.
Fabricated metal	n.a.	n.a.	0.3	1.5	1.3
Primary metal	n.a.	n.a.	0.2	0.5	0.8
Petroleum	0.09	0.45	0.7	0.8	1.0
Paper	0.06	0.17	0.3	0.7	0.7
Food	0.02	0.04	0.10	0.3	0.3
Forest products	0.01	0.04	0.03	n.a.	0.5
Leather	0.01	0.02	0.03	n.a.	n.a.
Textiles and apparel	0.01	0.02	0.07	n.a.	0.6

SOURCE: Y. Brozen, "Trends in Industrial Research and Development," *Journal of Business*, July 1960, Tables 1–3, and *Research and Development in Industry, 1961* (National Science Foundation, 1964).

[a] The 1957 and 1961 figures are not entirely comparable with the earlier ones. See the *Source*.

as a percentage of sales has been highest in the aircraft, electrical equipment, instrument, and chemical industries (Tables 1.4 and 1.5). Of course, this is

[9] For a description of the magnitude, nature, rationale, and history of the role of the federal government in research and development, see Mansfield, *Economics of Technological Change*, Chapter 6.

due in considerable part to the fact that these industries carry out a great deal of research and development for the federal government: In 1964, the federal government financed about 90 percent of this work in the aircraft industry; 60 percent in the electrical equipment industry; and 40 percent in the instruments industry. The situation in all industries in 1964 is shown in Table 1.5. When company-financed research and development rather than research and development performance is considered, the differences among industries are reduced, but the industries remain in much the same rank order—instruments, electrical equipment, chemicals, and machinery being highest.

Table 1.5. *R and D Performance and Amount Financed by Federal Government, by Industry, 1964 (millions of dollars)*

INDUSTRY	R AND D PERFORMANCE	AMOUNT FINANCED BY FEDERAL GOVERNMENT
Food and kindred products	135	n.a.
Paper and allied products	73	—
Chemicals and allied products	1,284	230
Industrial chemicals	856	172
Drugs and medicines	235	11
Other chemicals	193	47
Petroleum refining and extraction	337	27
Rubber products	150	26
Stone, clay, and glass products	133	10
Primary metals	191	8
Primary ferrous products	113	2
Nonferrous and other metal products	78	6
Fabricated metal products	152	18
Machinery	1,028	258
Electrical equipment and communication	2,635	1,628
Communication equipment and electronic components	1,480	973
Other electronic equipment	1,154	655
Motor vehicles and other transportation equipment	1,189	324
Aircraft and missiles	5,097	4,607
Professional and scientific instruments	483	208
Scientific and mechanical measuring instruments	210	120
Optical, surgical, photographic and other instruments	273	88
Textiles	32	2
Lumber	11	n.a.

SOURCE: *National Science Foundation Review of Data on Science Resources*, No. 7, Jan. 1966.

n.a. Not available.

As to the amount of basic research,[10] applied research[11] and development[12] performed by industry, Table 1.6 provides a breakdown for 1964.[13] Basic research constitutes the largest percentage of total research and development performance in the chemical and petroleum industries, but in no case does it exceed 20 percent of the total. Development is a particularly large percent of the total in the aircraft, electrical equipment, and machinery industries. For all industries combined, about 4 percent of the total was basic research, 20 percent was applied research, and 76 percent was development. Thus, the bulk of the funds goes for development, not research.

The McGraw-Hill survey of business plans for new plant and equipment provides further information regarding the character of the research and development being carried out by industry. In all manufacturing industries combined, about 47 percent of the firms reported in 1962 that their main purpose was to develop new products, 40 percent reported that it was to improve existing products, and 13 percent reported that it was to develop new processes. Development of new products seemed to be particularly important in the electrical equipment and the chemical and fabricated metal industries. Improvement of existing products seemed to be particularly important in the transportation equipment, machinery, auto, steel, and textile industries. Development of new processes was particularly important in the petroleum and rubber industries.[14]

We have seen that there are large interindustry differences in the ratio of company–financed research and development expenditures to sales. Why do these differences persist? First, industries differ considerably in the value their customers place on increased performance. Being second–best in product performance in some fields is not a great handicap because customers do not care very much about the difference in performance. On the other hand, in other industries, a second–best product has relatively little value. Second, industries may differ considerably in the ease with which research and

[10] According to the National Science Foundation, basic research includes "research projects which represent original investigation for the advancement of scientific knowledge and which do not have specific commercial objectives, although they may be in fields of present or potential interest to the reporting company." See National Science Foundation, *Methodology of Statistics on Research and Development* (Washington, 1959), p. 124.

[11] Applied research includes "research projects which represent investigation directed to discovery of new scientific knowledge and which have specific commercial objectives with respect to either products or processes. Note that this definition of applied research differs from the definition of basic research chiefly in terms of the objective of the reporting company." *Ibid.*, p. 124.

[12] Development includes "technical activity concerned with non-routine problems which are encountered in translating research findings or other general scientific knowledge into products or processes. It does not include routine technical services to customers or other items excluded from definitions of research and development." *Ibid.*

[13] Note that the distinctions between research and development and between development and more routine design improvements are blurred. See Mansfield, *The Economics of Technological Change* (New York: Norton, 1968), Chap. 3.

[14] These figures are several years old, but it seems doubtful that there has been much of a change since 1962.

development can bring about significant inventions. Third, industries differ in market structure. Industries composed of many small firms are unlikely

Table 1.6. *Distribution of Funds for the Performance of Basic Research, Applied Research, and Development, by Industry, 1964 (in percent)*

INDUSTRY	BASIC RESEARCH	APPLIED RESEARCH	DEVELOPMENT	TOTAL
Food and kindred products	9	47	44	100
Paper and allied products	3	36	62	100
Chemicals and allied products	13	n.a.	n.a.	100
Industrial chemicals	13	n.a.	n.a.	100
Drugs and medicines	16	49	35	100
Other chemicals	n.a.	23	68	100
Petroleum refining and extraction	15	45	39	100
Rubber products	7	20	73	100
Stone, clay, and glass products	5	35	59	100
Primary metals	6	37	57	100
Primary ferrous products	7	n.a.	n.a.	100
Nonferrous and other metal products	4	43	53	100
Fabricated metal products	3	23	75	100
Machinery	2	14	84	100
Electrical equipment and communication	5	14	81	100
Communication equipment and electronic components	8	16	76	100
Other electrical equipment	2	12	86	100
Motor vehicles and other transportation equipment	3	n.a.	n.a.	100
Aircraft and missiles	1	16	83	100
Professional and scientific instruments	n.a.	n.a.	77	100
Scientific and mechanical measuring instruments	3	11	86	100
Optical, surgical, photographic, and other instruments	n.a.	n.a.	n.a.	100
Textiles	3	50	47	100

SOURCE: *National Science Foundation Review of Data on Science Resources*, No. 7, Jan. 1966.

n.a. Not available.

to spend as much on research and development as somewhat less-fragmented industries. More will be said on this score in Chapter 2.[15]

[15] For a much more detailed discussion of the nature, organization, and management of industrial R and D, see Mansfield, *Economics of Technological Change*, Chap. 3.

7. Industrial Research and Development: Profitability and Uncertainty

Two of the most important characteristics of any economic activity are its profitability and the uncertainty involved. Research and development is no exception. Although it is extremely difficult to evaluate the returns from an investment in research and development, a firm must make the best estimates it can. For some years, the McGraw-Hill Economics Department gathered data from firms regarding the expected profitability of their research and development programs. Table 1.7 shows for each industry, in 1958 and 1961,

Table *1.7. Expected Average Pay-out Periods from R and D Expenditures, 1958 and 1961*

	PERCENT OF COMPANIES ANSWERING					
	1958			1961		
INDUSTRY	LESS THAN 3 YEARS	3 TO 5 YEARS	6 YEARS AND OVER	3 YEARS OR LESS	4 TO 5 YEARS	6 YEARS AND OVER
Iron and steel	50	50	0	38	50	12
Nonferrous metals	42	42	16	64	18	18
Machinery	49	45	6	51	39	10
Electrical machinery	23	69	8	61	32	7
Autos, trucks, and parts	40	60	0	54	40	6
Transportation equipment (aircraft, ships, railroad equipment)	24	65	11	43	44	13
Fabricated metals and instruments	24	71	5	77	14	9
Chemicals	15	56	29	33	41	26
Paper and pulp	25	69	6	50	32	18
Rubber	50	17	33	38	38	24
Stone, clay, and glass	44	50	6	38	46	16
Petroleum and coal products	12	63	25	17	33	50
Food and beverages	37	54	9	54	43	3
Textiles	65	29	6	76	24	0
Miscellaneous manufacturing	66	31	3	71	25	4
All manufacturing	39	52	9	55	34	11

SOURCE: McGraw-Hill, *Business Plans for Expenditures on Plant and Equipment* (annual).

the distribution of firms classified by their expected pay-out period for research and development. Although the pay-out period is a very crude measure of profitability, it is all that is available on a widespread basis. According to McGraw-Hill, the 1958 expected returns on research and development were "significantly better than the typical returns or pay-off, on investment in new plant and equipment, . . . [which helps to] make it clear why many companies with a given amount of capital to reinvest found it profitable to increase the proportion going to research and development." [16] In more recent years, there is considerable evidence that firms have been scrutinizing their research and development expenditures more carefully, and that expected returns have sometimes been adjusted downward.

One of the most obvious and important characteristics of research and development is the uncertainty of the outcome. Fundamentally, research and development is a learning process. Chance plays a crucial role, and a long string of failures must often be sustained before any sort of success is achieved. For example, a recent survey of 120 large companies doing a substantial amount of research and development indicates that in half of these firms at least 60 percent of the research and development projects never resulted in a commercially used product or process. (The smallest failure rate for any of these firms was 50 percent.) Moreover, even when a project resulted in a product or process that was used commercially, the profitability of its use was likely to be quite unpredictable [117].

A study carried out by The RAND Corporation [106] goes further in describing the extent of the difficulties in predicting the results of development projects. First, the study showed that there were substantial errors in the estimates (made prior to development) of the costs of producing various types of military hardware. When adjusted for unanticipated changes in factor prices and production lot sizes, the average ratio of the actual to estimated cost was 1.7 (fighters), 3.0 (bombers), 1.2 (cargoes and tankers), and 5.2 (missiles). Thus, the estimates made prior to the development of these types of equipment were off, on the average, by as much as 400 percent, and they almost always understated the true subsequent costs.

Second, the extent to which costs were understated was directly related to the extent of the technical advance. In cases where a "large" technical advance was required, the average ratio was 4.2; in cases where a "small" technical advance was required, the average ratio was 1.3. Moreover, when corrected for bias there was much more variation in the ratio in cases where the required technical advance was large than in those where it was small. Thus, as would be expected, the uncertainty was greater for more ambitious projects than for less ambitious ones.

Third, there were very substantial errors in the estimated length of time it would take to complete a project. For ten weapons systems the average error was 2 years, the maximum being 5 years. The average ratio of the actual to the expected length of time was 1.5, indicating once again that estimates tend to be overly optimistic. The results suggest too that the estimates are

[16] See D. Keezer, D. Greenwald, and R. Ulin, "The Outlook for Expenditures on Research and Development During the Next Decade," *Amer. Economic Rev.*, May 1960.

more accurate when "small" technical advances are required than when "large" technical advances must be made.

Fourth, given the extent of the technical advance that had to be made, the estimates of the development and production costs and the required development time became more accurate as the project ran its course. For example, at the early stages of projects requiring advances of "medium" difficulty the average ratio of the actual to expected cost was 2.15 and the standard deviation was 0.57. At the middle stages of such projects the average ratio was 1.32 and the standard deviation was 0.39. At the late stages of such projects the average ratio was 1.06 and the standard deviation was 0.18.

The findings of the RAND study pertain entirely to military research and development. To some extent, they reflect the fact that defense contractors have had an incentive to make optimistic estimates, the penalties for over-optimism often having been small and the possible rewards having been large. They also reflect the ambitious nature of development in the defense sphere. Nonetheless, they are not an utterly undependable guide to the civilian economy. In Chapter 3 there is evidence suggesting that the errors in estimation in the civilian economy are also quite large, although smaller than those presented above.

8. Summary and Conclusions

The principal conclusions of this chapter are as follows: First, technological change results in a change in the production function or in the availability of new products. Since there is no satisfactory way to measure the rate of technological change directly, economists often measure it by its effects. The rate of technological change in an industry depends to a large extent on the amount of resources devoted by firms, by independent inventors, and by government to the improvement of the industry's technology. The amount of resources devoted by the government depends on how closely the industry is related to the defense, health, and other social needs for which the government assumes major responsibility, on the extent of the external economies generated by the relevant research and development, and on more purely political factors. Since the amount of resources devoted by industry and independent inventors depends heavily on the anticipated profitability of their use, the rate of technological change depends on factors which influence the rewards from particular kinds of technological change and on factors which influence the cost of making particular kinds of technological change. Among the former factors are the nature of product demand and the configuration of input prices; among the latter factors are advances in basic science.

Second, besides being influenced by the quantity of resources an industry devotes to improving its own technology, an industry's rate of technological change depends on the quantity of resources devoted by other industries to the improvement of the capital goods and other inputs it uses. In addition, there is another kind of interdependence among industries, technology meant for one industry turning out to be useful for others as well. An industry's rate of technological change also depends on its market structure; the legal arrangements under which it operates; the attitudes toward technological change of management, workers, and the public, and the organization and management of its research and development.

Third, existing measures of the rate of technological change suffer from important limitations. However, using any of these measures there seems to have been a considerable increase in the level of technology in the United States throughout this century. For example, the total productivity index increased by about 1.7 percent per year in the private economy. The rate of increase varied considerably among industries and over time. Over the long run it seems to have been higher in communications and transportation than in mining, manufacturing, and farming. Within manufacturing it seems to have been highest in rubber, transportation equipment, tobacco, chemicals, printing, glass, fabricated metals, textiles, and petroleum. In a comparison of the United States with Germany, Japan, Canada, and the United Kingdom during the postwar period, our rate of productivity growth seemed lower than that of Germany and Japan, but at least equal to that of Canada and the United Kingdom.

Fourth, total research and development expenditures in the United States have increased spectacularly during the last several decades, much of the research and development performed by industry being financed by the federal government. Three departments and agencies—the Department of Defense, the National Aeronautics and Space Administration, and the Atomic Energy Commission—account for about 90 percent of the research and development expenditures of the federal government. The primary purpose of the expenditures made by these agencies is to develop and improve weapons systems, to push forward the nation's space program, and to develop new applications for atomic energy. The bulk of their expenditures is for development, not research, and the bulk of their research expenditures is in the physical and engineering sciences.

Fifth, the performance of research and development is spread very unevenly among industries; as a percent of sales it is highest in the aircraft, electrical equipment, instrument, and chemical industries. When company-financed research and development rather than research and development performance is considered, the differences among industries are reduced, but the industries remain in much the same rank order. These interindustry differences may reflect differences among industries in the profitability of research and development. However, this hypothesis is difficult to test because of the problems in measuring the returns from research and development. More will be said about these problems in Part II.

Part II

Industrial Research and Development

CHAPTER 2

Industrial Research and Development Expenditures: Determinants and Relation to Size of Firm and Inventive Output

This chapter is concerned with the research and development expenditures of private industry, primary attention being devoted to the following three questions:[1] First, what determines the level of a firm's privately financed research and development expenditures? Since most of the R and D carried out by industries other than aircraft, electrical equipment, and instruments is financed privately, it obviously is important that we explore the determinants of such expenditures.[2] Second, what is the relationship between a firm's research and development expenditures and various crude measures of inventive output? It is important to determine whether these expenditures are closely related to inventive output, and if so whether there seem to be marked economies of scale with respect to various kinds of research and development. It is also interesting to determine the effects of a firm's size on the productivity of its research and development

[1] Some relevant studies in this area are Y. Brozen, "Invention, Innovation, and Imitation," *Amer. Economic Rev.* (May 1951); M. Gainsbrough, "Allocation of Resources to Research and Development," *Conference on Research and Development* (National Science Foundation, 1958); I. Horowitz, "Regression Models for Company Expenditures on and Returns from Research and Development," *IRE Transactions on Engineering Management*, 1960; J. Minasian, "The Economics of Research and Development," *The Rate and Direction of Inventive Activity* (Princeton, 1962); J. Quinn, "Long-Range Planning of Industrial Research," *Harvard Business Review*, 1961; and A. Rubenstein, "Setting Criteria for R and D," *Harvard Business Review* (Jan. 1957).

[2] Like the studies cited in note 1, we lump together R and D expenditures throughout this chapter. For a discussion of the determinants and effects of publicly financed R and D expenditures, see E. Mansfield, *The Economics of Technological Change* (New York: Norton, 1968), Chaps. 3 and 6.

expenditures. Third, do the largest firms in various industries spend more on research and development relative to their size, than firms one-fifth or one-tenth of their size? As pointed out in Chapter 1 this question is of considerable importance to economists interested in industrial organization, since it is often maintained that giants like DuPont and Standard Oil (N.J.) are required to promote technical progress.

In section 1, a simple model is presented to explain the level of a firm's research and development expenditures in an industry in which government financing plays a relatively small role—like chemicals, petroleum, drugs, glass, and steel. In section 2, this model is tested, and several implications of the empirical results are discussed. In sections 3 and 4, we describe the 1945–1958 behavior of research and development expenditures for thirty-five firms and present some rough forecasts of their later behavior. The relationship between a firm's size and its research and development expenditures is analyzed in section 5, and the relationship between a firm's research and development expenditures and the number of significant inventions or innovations it produced is examined in section 6.

1. The Level of a Firm's Research and Development Expenditures: A Simple Model

As a first step, we assume that a firm's managers have in mind some desired or target level of R and D expenditures which they would aim for if adjustments in personnel and plant could be made instantaneously and if the inefficiencies involved in too rapid a change in R and D expenditures could be avoided.[3] When they determine the R and D budget for the coming year, we assume that they take this year's actual R and D expenditures as a base from which to figure and set next year's R and D budget so as to move a certain fraction of the way from this base toward the desired level.[4] Thus, if $r_i(t)$ is the ith firm's R and D budget for year t, $\tilde{R}_i(t)$ is its desired expenditure for year t, and $R_i(t-1)$ is its actual expenditure in year $(t-1)$,

$$r_i(t) = R_i(t-1) + \theta_i(t)[\tilde{R}_i(t) - R_i(t-1)], \tag{1}$$

[3] I presented a preliminary version of this model in my "Comment," *The Rate and Direction of Inventive Activity* (Princeton, 1962). Although I pointed out that $\tilde{R}_i(t)/S_i(t)$ and $\theta_i(t)$, both of which are defined below, might vary over time, I assumed for simplicity that they were constant in the preliminary tests regarding chemical firms reported there. For an application of these results, see Y. Brozen, "The Future of Industrial Research," *Journal of Business*, 1961.

[4] When it is assumed that $\tilde{R}_i(t) > R_i(t-1)$, which typically seems to have been the case in the postwar period, three reasons become apparent as to why the firm moves only part way toward $\tilde{R}_i(t)$: First, it takes time to hire people and build laboratories. Second, there are often substantial costs involved in expanding $R_i(t)$ too rapidly, because it is difficult to assimilate large percentage increases in R and D staff. The costs of changing $R_i(t)$ are not included in determining $\tilde{R}_i(t)$. Third, the firm may be uncertain as to how long expenditures of $R_i(t)$ can be maintained. It does not want to begin projects that will soon have to be interrupted.

where $\theta_i(t)$ is the fraction of the way the ith firm moves toward the desired expenditure. If it is assumed that the actual expenditure in year t equals the budgeted amount plus a random error term,

$$(2) \qquad R_i(t) = R_i(t-1) + \theta_i(t)[\tilde{R}_i(t) - R_i(t-1)] + z_i(t),$$

where $z_i(t)$ is a random variable with zero expected value.[5]

In equation (2) there are two variables, $\tilde{R}_i(t)$ and $\theta_i(t)$, that must be explained. What determines $\tilde{R}_i(t)$? Suppose that the ith firm, when planning the extent of its research activities in year t, lists all proposed R and D projects and makes a rough estimate of the expected rate of return from each one.[6] Let this estimated rate of return be ρ and assume that for proposed R and D projects with relatively high values of ρ the probability distribution of ρ, like that of many other economic variables, can be approximated by the Pareto law.[7] Thus, the probability that a proposed R and D project has a ρ exceeding x is

$$(3) \qquad C_i(t)x^{-U_i(t)}.$$

Letting r be the minimum rate of return required by firms in the industry to justify a risk-free investment, assume that the probability that ρ exceeds r equals $H_i(t)$. Thus,

$$(4) \qquad Pr\ \{\rho > x\} = H_i(t)\left(\frac{x}{r}\right)^{-U_i(t)}$$

If the ith firm could make the necessary adjustments instantaneously and without inefficiencies stemming from rapid expansion, it would accept all R and D projects exceeding a certain minimum expected rate of return, ρ^*, which (like r) is assumed to be the same for all firms in the industry. Consequently, the probability that a proposed R and D project will be accepted is

[5] This assumes that, on the average, the budgeted and actual expenditures are equal. Judging from the interview data described in Appendix A, this seems to be a reasonable approximation, but if for some reason the expected value of $z_i(t)$ is nonzero, the only effect it has on the model is that equation (2) will contain an intercept. This means that equation (14) will, too.

[6] Of course, the rate of return on the investment in a particular R and D project is notoriously difficult to measure (Cf. Rubenstein, "Setting Criteria"), and I do not assume that the firm's estimates are very accurate. I assume only that the firm makes a rough estimate—formal or informal—of a project's prospective profitability, allowing in some way for risk. This assumption seems to be borne out by Gainsbrough (in "Allocation of Resources"), by Quinn (in "Long-Range Planning") and by others, as well as by the interview data in Appendix A.

[7] The usefulness of the Pareto distribution in representing the upper tail of various distributions in economics is well known. (See B. Mandelbrot, "New Methods in Statistical Economics," *Journal of Political Economy* [Oct. 1963].) Needless to say, the choice of this distribution is to some extent a matter of convenience; a number of others might have done about as well. For simplicity, I assume that the cost of all R and D projects is the same. This assumption could easily be relaxed without altering the results.

$$H_i(t)\left(\frac{\rho^*}{r}\right)^{-U_i(t)}.$$

If the total cost of all R and D projects that are proposed is

(5) $$M_i(t) = hS_i(t)^{\nu_1}k_i(t),$$

where h varies from industry to industry, ν_1 is positive, $S_i(t)$ is the ith firm's sales in year t, and $k_i(t)$ is a random error term,[8] we have

(6) $$\tilde{R}_i(t) = hH_i(t)S_i^{\nu_1}(t)\left(\frac{\rho^*}{r}\right)^{-U_i(t)}k_i(t).$$

Finally, if $\bar{\rho}_i(t)$ is the average expected rate of return from R and D projects that are accepted, it can be shown that $U_i(t) = \bar{\rho}_i(t)/[\bar{\rho}_i(t) - \rho^*]$. Thus,

(7) $$\tilde{R}_i(t) = hHS_i^{\nu_1}(t)\left(\frac{\rho^*}{r}\right)^{\frac{\bar{\rho}_i(t)}{\rho^*-\bar{\rho}_i(t)}}\frac{H_i(t)}{H}k_i(t),$$

where H is the average value of $H_i(t)$ in the industry; and

(8) $$ln\ \tilde{R}_i(t) = ln\ \nu_o + \nu_1\ ln\ S_i(t) + \nu_2\frac{\bar{\rho}_i(t)}{\rho^* - \bar{\rho}_i(t)} + ln\left[k_i(t)\frac{H_i(t)}{H}\right],$$

where $\nu_o = hH$, $\nu_2 = ln\ (\rho^*/r)$, and $ln\left[k_i(t)\frac{H_i(t)}{H}\right]$ is an error term.[9]

What determines $\theta_i(t)$? First, one would expect $\theta_i(t)$ to be inversely related to $[\tilde{R}_i(t) - R_i(t-1)]/R_i(t-1)$. If attaining its desired level of expenditures implies that a firm's R and D activities must be increased by a very large percentage, the firm will move a smaller proportion of the way toward this

[8] It may be objected that r is so small that the Pareto distribution is unlikely to provide a good approximation to $\Pr\{\rho > r\}$. But it should be noted that a great many proposals are made for R and D projects, and only a small proportion are estimated to yield a non-negative rate of return. Thus, r is quite far out in the tail of the distribution, and the Pareto law is likely to provide a good approximation. Large firms would be expected to generate more R and D proposals than small firms for the following reasons: They have more workers; many projects would be worthwhile to a large firm but not a small one (because the saving per unit is spread over a larger volume, and the large firm has the capital to exploit things which are beyond the reach of the small firm); and the risks would often be less for the large firm. Sales data are used as a measure of firm size throughout this chapter. If other measures (assets, employment) were used, it seems doubtful that the results would differ appreciably.

Note too that we could easily make the distribution of rates of return from R and D projects a function of the level of a firm's sales. This would change nothing of consequence. Thus there is no contradiction between the model used here and the one in Chapter 4, which hypothesizes that the returns from R and D are dependent on a firm's sales.

[9] I assume that $U_i(t) > 1$, which is consistent with the results in Table 2.1. Of course, $\bar{\rho}_i(t)$ depends on the actual profitability of proposed R and D projects, and H varies from industry to industry. Equation (8) assumes that $H_i(t)/H$ is statistically independent of the exogenous variables in the equation.

desired level than if it implies only a small percentage increase in its R and D activities. There are considerable costs involved in a very rapid expansion of a firm's R and D department, the importance of which was stressed in interviews with various executives in the industries described below.[10] Of course, we assume that $\tilde{R}_i(t) > R_i(t-1)$, but this almost always seems to have been the case during the postwar period. Whether or not it will typically be the case in the future is not obvious, but the model could easily be altered to accommodate periods when $\tilde{R}_i(t) < R_i(t-1)$.

Second, one would expect $\theta_i(t)$ to be directly related to the ratio of the ith firm's profits to its R and D expenditures in year $(t-1)$. When $[\tilde{R}_i(t) - R_i(t-1)]/R_i(t-1)$ is held constant, this factor is directly related to the percent of a firm's profits in year $(t-1)$ that would have been absorbed by an increase in R and D expenditures to the desired level. If a large percent of a firm's profits would have been absorbed by such an increase, the firm will tend to be more cautious in moving toward the desired level. Conversely, if a firm spent less of its profits on R and D than others in its industry during year $(t-1)$, it may move more rapidly toward its desired level in an effort to catch up with its competitors. This factor was also stressed in the interviews, and we assume once again that $\tilde{R}_i(t) > R_i(t-1)$.[11]

Let $P_i(t-1)$ be the ratio of the ith firm's profits to its R and D expenditures in year $(t-1)$ and $\pi_i(t-1)$ be the ratio of $P_i(t-1)$ to the average value of $P_i(t-1)$ in the industry, and assume that

$$\theta_i(t) = \nu_3 + \nu_4 \frac{R_i(t-1)}{[\tilde{R}_i(t) - R_i(t-1)]} + \nu_5 \pi_i(t-1) + k_i''(t), \qquad (9)$$

where ν_4 and ν_5 are positive and $k_i''(t)$ is a random error term. Of course, equation (9) is assumed to hold only within a certain range—presumably the relevant one. We would expect ν_3 to differ among industries, allowing for such factors as the extent to which engineering and scientific personnel are in relatively short supply.[12]

[10] Because of the difficulties in obtaining and assimilating a large increase in staff and equipment, some of these firms rule out annual increases in R and D expenditures exceeding 10 to 15 percent.

[11] Let $P_i(t-1)$ be the ratio of the ith firm's profits to its R and D expenditures in year $(t-1)$. Since the proportion of the ith firm's profits in year $(t-1)$ that would be absorbed by an increase in R and D expenditures to the desired level is the ratio of $[\tilde{R}_i(t) - R_i(t-1)]/R_i(t-1)$ to $P_i(t-1)$, it follows that this proportion is inversely related to $P_i(t-1)$ when $[\tilde{R}_i(t) - R_i(t-1)]/R_i(t-1)$ is held constant. If a firm spent a relatively large proportion of its profits on R and D in year $(t-1)$, it seems likely to move more slowly in its plans for year t because of liquidity constraints and because it may believe it is getting out of line with its competitors. The importance of $P_i(t)$ was stressed in practically all of the interviews. Of course, cash flow might have been used rather than profits, but the results would probably have been much the same. In equation (9) we use as an independent variable the ratio of $P_i(t-1)$ to the average value of $P_i(t-1)$ in the industry. Since a "large" and "small" value of $P_i(t-1)$ vary from industry to industry, this normalization seems reasonable. Also, the value of $P_i(t-1)$ relative to the past may be important.

[12] The form of equation (9) was chosen primarily for its convenient properties and its simplicity. Note, however, that because $0 \leq \theta_i(t) \leq 1$, equation (9) can hold only for values

Taken together, equations (2), (8), and (9) constitute a simple model to explain the level of the ith firm's R and D expenditures during year t, the exogenous variables being $R_i(t-1)$, $S_i(t)$, $\frac{\bar{p}_i(t)}{\rho^* - \bar{p}_i(t)}$, and $\pi_i(t-1)$.[13] Of course, this model is oversimplified in many respects. To keep it manageable and operational, we restrict ourselves to only a few exogenous variables that can be measured at least roughly. Moreover, the model is designed primarily to explain differences among firms in a given industry rather than inter-industry differences. To explain interindustry differences, more attention would have to be paid to the factors influencing h. Finally, the model is a better representation of decision-making regarding development and applied research than basic research; but this shouldn't be too important since basic research is a very small component of industrial R and D.

2. Tests of the Model and Implications of the Findings

To test the model, we estimate the parameters in equations (8) and (9) for the chemical and petroleum industries, test whether they have the hypothesized signs, and see how well these equations fit the data. To obtain these results, we use estimates of $\theta_i(t)$, $\bar{p}_i(t)$, ρ^* and $\tilde{R}_i(t)$ derived from interviews and correspondence with eight major firms in these industries. Appendix A describes in detail how these data were obtained. Although the number of firms is quite small and the data are rough, the resulting estimates should be of considerable use in testing the model.

First, consider the parameters in equation (8). Using these interview data regarding $\tilde{R}_i(t)$, ρ^*, and $\bar{p}_i(t)$ for eight firms in 1958 and five firms in 1962, and using corresponding data from *Moody's* regarding $S_i(t)$, we obtain least-squares estimates of ν_o, ν_1, and ν_2, the resulting regression equation being

$$(10) \qquad ln\ \tilde{R}_i(t) = \begin{Bmatrix} -2.38 \\ -0.22 \end{Bmatrix} + \underset{(0.06)}{0.89}\ ln\ S_i(t) + \underset{(0.15)}{0.75}\bar{p}_i(t)/[\rho^* - \bar{p}_i(t)],$$

where the top figure in brackets pertains to the petroleum industry and the bottom figure pertains to chemicals. (See Table 2.1 for the basic data.) All

---

of $R_i(t-1)/[\tilde{R}_i(t) - R_i(t-1)]$ and of $\pi_i(t-1)$ in a certain range. I assume in the following section that it holds for $R_i(t-1)/[\tilde{R}_i(t) - R_i(t-1)] < 40$; in sections 3 and 4, I assume that it always holds. See note 26.

For some discussion of the market for engineers and scientists, see D. Blank and G. Stigler, *The Demand and Supply of Scientific Personnel* (National Bureau of Economic Research, 1957) and K. Arrow and W. Capron, "Dynamic Shortages and Price Rises: The Engineer-Scientist Case," *Quarterly Journal of Economics* (May 1959).

[13] I assume that the model will be used only for relatively short-run purposes. If not, one must take account of variation over time in the ν's. All of the exogenous variables other than $S_i(t)$ can be measured at time $t-1$. If it seems desirable for all such variables to be measurable at time $t-1$, the firm's forecast of sales in year t or its actual sales in year $t-1$ probably can be used throughout instead of $S_i(t)$.

Table 2.1. *Values of Variables in Equations (8) and (9), Eight Firms, Chemical and Petroleum Industries*

FIRM	$\tilde{R}_i(t)/S_i(t)$	$\bar{\rho}_i(t)/[\rho^* - \bar{\rho}_i(t)]$[a]	$\theta_i(t)$	$R_i(t-1)/[\tilde{R}_i(t) - R_i(t-1)]$	$\pi_i(t-1)$
1962					
C1	0.0297	−3.42	0.73	4.9	1.0
C2	.0505	−3.00	0.94	35.4	0.7
C3	.0260	−3.86	1.00	>40	1.3
P1	.0074	−2.00	1.00	>40	1.4
P2	.0090	−1.67	1.00	34.7	0.6
1958					
C1	.0283	−3.50	0.23	3.6	0.8
C2	.0455	−3.00	0.75	18.9	0.9
C3	.0262	−3.50	1.00	5.6	1.2
P1	.0082	−2.00	1.00	8.5	1.1
P2	.0088	−2.50	0.20	4.2	0.7
P3	.0048	−2.50	1.00	7.6	1.3
P4	.0165	−1.50	—	—	—
P5	.0126	−2.00	0.10	1.1	0.9
1955					
P2	.0068	—	0.45	3.2	0.9
P3	.0072	—	0.58	7.3	1.2

SOURCE: See Appendix A.

SYMBOLS: $\tilde{R}_i(t)/S_i(t)$ is the ratio of the desired R and D expenditures of the ith firm during year t to its sales during year t; ρ^* is the minimum expected rate of return from R and D projects that are accepted; $\bar{\rho}_i(t)$ is the average expected rate of return from R and D projects accepted by the ith firm during year t; $\theta_i(t)$ equals $[r_i(t) - R_i(t-1)]/[\tilde{R}_i(t) - R_i(t-1)]$; $R_i(t)$ is the actual R and D expenditures of the ith firm during year t; and $\pi_i(t)$ equals $P_i(t)/P(t)$, where $P_i(t)$ is the ratio of the ith firm's profits to its R and D expenditures in year t and $P(t)$ is the average value of $P_i(t)$ in the industry during year t.

NOTE: Most of the data were obtained with the understanding that they would remain confidential. Thus the names and sales of the companies are not given, C1 standing for the first chemical firm, P1 for the first petroleum firm, etc. For Company 1 in the chemical industry, the data refer to 1959, not 1958. For Company 4 in the petroleum industry, $\tilde{R}_i(t) < R_i(t)$; thus, the model regarding $\theta_i(t)$ does not apply and the last three columns of the table are left blank. No estimates of $\bar{\rho}_i(t)/[\rho^* - \bar{\rho}_i(t)]$ are available for 1955.

[a] This ratio often is an integer or an integer plus one half, because the respondents usually rounded the payout periods (or other measures) used to estimate $\bar{\rho}_i(t)$ and ρ^* to the nearest integer.

FIGURE 2.1. Actual and calculated values of $ln\ \tilde{R}_i(t)$, eight petroleum and chemical companies, 1958 and 1962.

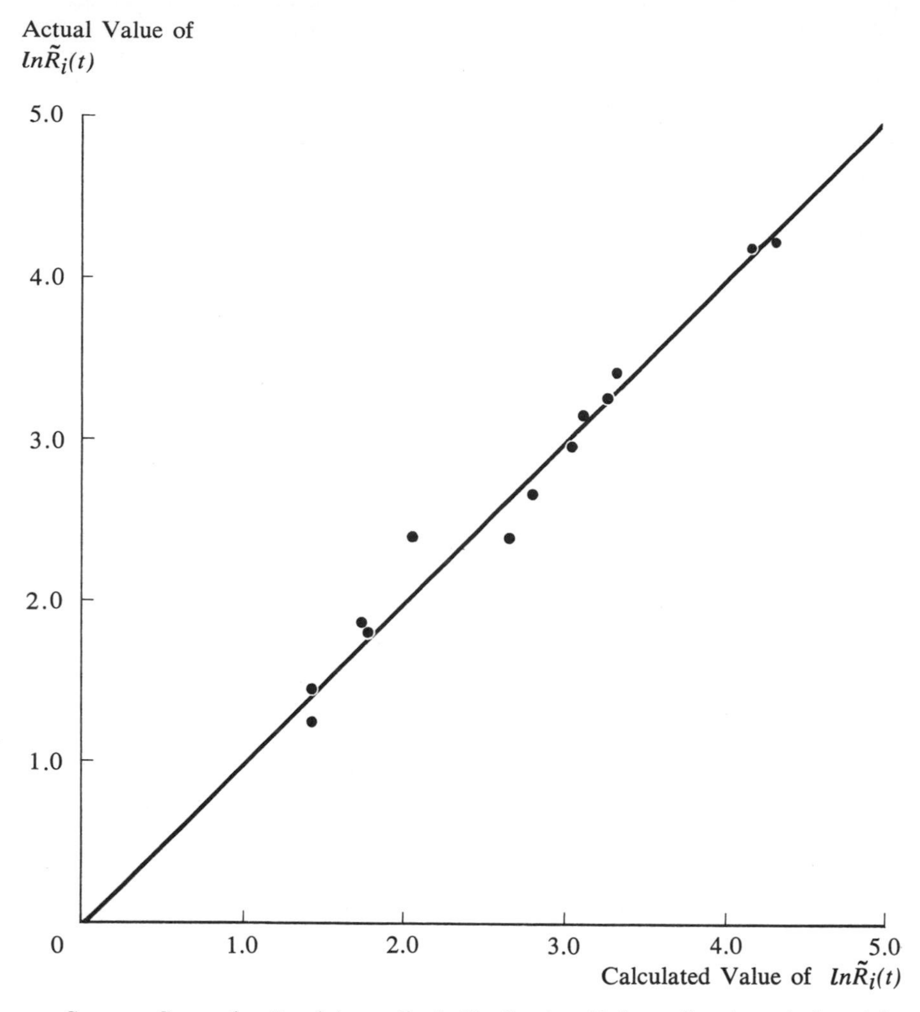

SOURCE: See section 2 and Appendix A. The line is a 45-degree line through the origin.

NOTE: $\tilde{R}_i(t)$ is the ith firm's desired R and D expenditures in year t (in $ millions).

of the regression coefficients are highly significant. Figure 2.1 shows that equation (10) fits the data very well, the coefficient of correlation (adjusted for degrees of freedom) being 0.97.[14]

Second, consider the parameters in equation (9). Using 1955, 1958, and 1962 data regarding $\tilde{R}_i(t)$ and $R_i(t-1)$ obtained from interviews and corre-

[14] Note four things: First, I added another term representing interyear differences in ν_0 to equation (10), but found that it was statistically nonsignificant. Second, the estimate of ν_2 implies that ρ^* is about twice as big as r, which seems reasonable because of the uncertainty of the outcomes of R and D projects. Third, errors in measuring $\bar{\rho}_i(t)/[\rho^* - \bar{\rho}_i(t)]$ would be likely to bias the estimate of ν_2 toward zero. Fourth, if more data were available, it would be useful to test whether ν_1 and ν_2 differ among industries.

spondence with these firms and corresponding data regarding $\pi_i(t-1)$ from *Moody's*, we obtain least-squares estimates of ν_3, ν_4, and ν_5. Since the estimates of ν_3 do not differ significantly between industries, we assume that they were the same and find that

$$\text{(11)} \qquad \theta_i(t) = -0.95 + \underset{(0.005)}{0.029} \frac{R_i(t-1)}{[\tilde{R}_i(t) - R_i(t-1)]} + \underset{(0.25)}{1.38}\pi_i(t-1).$$

Figure 2.2 shows that equation (11) fits the data quite well, the correlation

FIGURE 2.2. Actual and calculated values of $\theta_i(t)$, seven petroleum and chemical companies, 1955, 1958, and 1962.

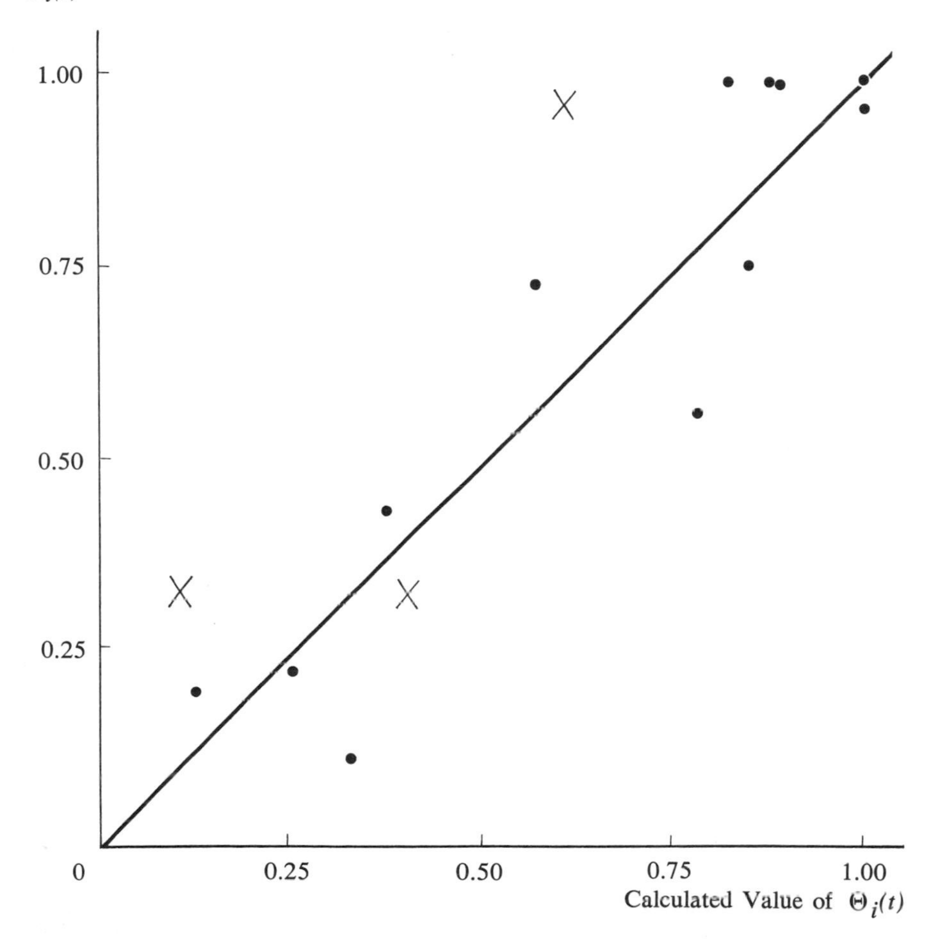

SOURCE: See section 2 and Appendix A. The line is a 45-degree line through the origin.

NOTE SYMBOLS: $\theta_i(t)$ equals $[r_i(t) - R_i(t-1)]/[\tilde{R}_i(t) - R_i(t-1)]$. The dots represent the data in Table 2.1 on which equation (11) was based. The X's represent "predictions" for other years for firms where the necessary data could be obtained. Note that a computed value of $\theta_i(t)$ exceeding one was set equal to one, because $0 \leq \theta_i(t) \leq 1$.

coefficient being 0.89. The estimates of ν_4 and ν_5 have the expected signs and are statistically significant. Moreover, equation (11) seems to be reasonably effective (considering the sampling errors in ν_3, ν_4, and ν_5) in "predicting" the values of $\theta_i(t)$ for these firms in a few additional years for which we could obtain data. The average error of the "predictions" (also shown in Figure 2.2) is about 0.20.[15]

The results of these tests are encouraging. However, it must be emphasized that the estimates of $\tilde{R}_i(t)$, $\bar{\rho}_i(t)$, and ρ^* are so rough and the number of observations are so small that the findings are only tentative. Nonetheless, this should not prevent us from examining the implications of the findings. For example, what would be the effect, in *money* terms, on a firm's expenditures of a change in the tax laws that increased the prospective profitability of each of its R and D projects by 100 δ percent?[16] If it is assumed that the model holds and that the firm's actual and desired expenditures would be approximately equal, the firm's spending on R and D under existing circumstances would be

$$\tilde{R}_i^o(t) = hHS_i(t)^{\nu_1}\left(\frac{\rho^*}{r}\right)^{\frac{\bar{\rho}_i(t)}{\rho^*-\bar{\rho}_i(t)}} k_i(t).$$

If there were a 100 δ percent increase in the value of ρ for all projects, its expenditures would be

$$\begin{aligned}\tilde{R}_i'(t) &= hS_i(t)^{\nu_1}k_i(t)[Pr\ \{(1+\delta)\rho > \rho^*\}]\\ &= hHS_i(t)^{\nu_1}\left(\frac{\rho^*}{(1+\delta)r}\right)^{\frac{\bar{\rho}_i(t)}{\rho^*-\bar{\rho}_i(t)}} k_i(t),\end{aligned}$$

the relative increase in its R and D expenditure being

$$\begin{aligned}[\tilde{R}_i''(t) - \tilde{R}_i^o(t)]/\tilde{R}_i''(t) &= \left[(1+\delta)^{\frac{\bar{\rho}_i(t)}{\bar{\rho}_i(t)-\rho^*}} - 1\right]\\ &\doteqdot \bar{\rho}_i(t)\delta/(\bar{\rho}_i(t)-\rho^*).\end{aligned} \tag{12}$$

Thus, if a change in policy resulted in a 1 percent increase in the prospective

[15] Note five things: First, if $R_i(t-1)/[\tilde{R}_i(t) - R_i(t-1)]$ exceeds 40, I assume that equation (9) does not hold and that $\theta_i(t) = 1$. (See note 12). This is arbitrary but reasonable. Second, dummy variables representing differences among years were added to equation (11) but were statistically nonsignificant. Third, the data regarding $\tilde{R}_i(t)$ used in the "predictions" are described in Appendix A. Unfortunately, since corresponding values of $\bar{\rho}_i(t)/[\rho^* - \bar{\rho}_i(t)]$ could not be obtained, they could not be used in connection with equation (10). They generally pertain to 1952 or 1955. Fourth, one reason why ν_3 does not differ significantly among industries or years may be that the labor market was about as tight in one industry or year as in another. Fifth, somewhat better estimates might have been obtained in equation (11) if the estimate of $\tilde{R}_i(t)$ from equation (10) had been used rather than the actual values of $\tilde{R}_i(t)$, but the differences would have undoubtedly been slight.

[16] The ad hoc Advisory Group on Science Policy of OECD claimed several years ago that there "is often a need to encourage increased industrial expenditures on research by means of tax or other incentives." (*Science and the Policies of Government*, 1963, p. 22). If so, one is faced with the problem of estimating the effects of various incentives, which is the topic discussed above.

profitability of R and D projects, R and D expenditures would increase by about 2 percent among the petroleum firms and by about $3\frac{1}{2}$ percent among the chemical firms.[17]

Finally, two additional points should be noted concerning the results. First, the fact that equation (10) fits so well seems to imply that the process by which a firm's R and D expenditures are determined is not so divorced from profit considerations as some observers have claimed. If firms "establish research laboratories without any clearly defined idea of what the laboratories could perform" [125, p. 20], and blindly devote some arbitrarily determined percentage of sales to R and D, it is difficult to see why equation (10) fits so well.[18]

Second, the fact that the estimate of ν_1 is significantly less than one means that among large petroleum and chemical firms increases in size result in no more than proportional increases in the total dollar volume of R and D proposals. Consequently, if $\bar{p}_i(t)$ is independent of a firm's size (in this range), one would expect larger firms to spend no greater proportion of their sales on R and D than smaller firms (in this range). This hypothesis is tested in section 5.

3. Estimates of Actual and Desired Research and Development Expenditures in the Chemical, Petroleum, Drug, Steel, and Glass Industries, 1945–1958

Data regarding R and D expenditures, particularly pre-1953 expenditures, are difficult to obtain. This section describes how much thirty-five major firms in five important industries spent annually on R and D during 1945–1958. In addition, rough estimates are made of the firms' desired R and D expenditures during 1945–1958, such estimates being useful because they indicate the equilibrium values toward which the firms were moving. Following customary procedure, both actual and desired R and D expenditures are expressed as a proportion of sales.

[17] Note three things: First, no procedure is given here to estimate the effect of a given change in policy on the firm's estimates of the profitability of R and D. We provide only an estimate of the effect on a firm's R and D expenditures, given that the former sort of estimate has been made. Second, for an increase in R and D expenditures to be entirely "real," the firm would have to face a perfectly elastic supply of factors. Our results can be used to determine a demand function for "real" R and D, but we also need a supply function to estimate the effect of a change in policy on the industry's real R and D expenditures. Third, some changes in policy may affect ρ^* as well as $\bar{p}_i(t)$.

[18] There has been a tendency in some quarters to emphasize the importance of non-economic motives and "fashion," rather than expectation of profit as determinants of the firm's spending on R and D. (See J. Jewkes, D. Sawers, and R. Stillerman, *The Sources of Invention* [New York: St. Martin's Press, 1958], p. 151; and Organization for European Economic Cooperation, *The Organization of Applied Research in Europe, the United States, and Canada*, OEEC, 1954, p. 29.) Such factors do operate but they may not be as important as has been claimed. Of course, our results indicate only that firms act in accord with their expectations of profit from R and D. No doubt, these expectations are affected by the expectations and decisions of their competitors.

The data regarding $R_i(t)/S_i(t)$ come primarily from correspondence with the firms and from Langenhagen [72].[19] Since it is difficult to obtain direct estimates of $\tilde{R}_i(t)/S_i(t)$, like those in Table 2.1, two rather bold assumptions are made in order to obtain indirect estimates. First, we assume that, during 1945–1958, a firm's desired R and D expenditures, as a percent of its sales, were a quadratic function of time with random disturbances, i.e.,

$$\tilde{R}_i(t)/S_i(t) = \alpha_{i1} + \alpha_{i2}t + \alpha_{i3}t^2 + u_i(t), \tag{13}$$

where $u_i(t)$ is a random error term, and time is measured in years from 1945.[20] Second, letting π_i be the average value of $\pi_i(t-1)$ for the ith firm during 1945–1958, it follows from equations (2) and (9) that

$$\begin{aligned} R_i(t) = {} & R_i(t-1) + \left[\nu_3 + \nu_4 \frac{R_i(t-1)}{[\tilde{R}_i(t) - R_i(t-1)]} + \nu_5\pi_i(t-1) \right. \\ & \left. + k_i''(t)\right][\tilde{R}_i(t) - R_i(t-1)] + z_i(t), \\ = {} & (1+\nu_4)R_i(t-1) + [\nu_3 + \nu_5\pi_i][\tilde{R}_i(t) - R_i(t-1)] \\ & + \{\nu_5[\pi_i(t-1) - \pi_i] + k_i''(t)\}[\tilde{R}_i(t) - R_i(t-1)] \\ & + z_i(t); \end{aligned} \tag{14}$$

and we assume that the last two terms on the right-hand side of equation (14) plus $u_i(t)S_i(t)[\nu_3 + \nu_5\pi_i]$ can be treated as a random error term (with zero expected value) which is independent of $R_i(t-1)$, $S_i(t)$, t, and t^2. Although these two assumptions are rough, they seem to be useful: the resulting estimates of $\tilde{R}_i(t)/S_i(t)$ agree reasonably well with those in Table 2.1. (See Appendix A.)

Given these assumptions, we estimate $\tilde{R}_i(t)/S_i(t)$ in two steps. Combining equations (13) and (14) and bearing in mind the second assumption,

$$R_i(t) = \beta_{i1}R_i(t-1) + \beta_{i2}S_i(t) + \beta_{i3}tS_i(t) + \beta_{i4}t^2S_i(t) + z_i''(t), \tag{15}$$

where $\beta_{i1} = (1 + \nu_4 - \nu_3 - \nu_5\pi_i)$, $\beta_{i2} = \alpha_{i1}(\nu_3 + \nu_5\pi_i)$, $\beta_{i3} = \alpha_{i2}(\nu_3 + \nu_5\pi_i)$, $\beta_{i4} = \alpha_{i3}(\nu_3 + \nu_5\pi_i)$, and $z_i''(t)$ is a random error term. The first step in the estimation procedure is to estimate the β's. The results, shown in Tables 2.2 and 2.3, indicate that almost all of the estimates of β_{i1} and β_{i2} have the expected signs and are of the expected orders of magnitude. Despite relatively

[19] The firms from which I obtained data were promised that their names would not be published. No attempt was made to obtain data regarding $\bar{\rho}_i(t)$ or direct estimates of $\tilde{R}_i(t)$, because it seemed likely that the nonresponse rate would increase markedly if such questions were added.

[20] Although this assumption is somewhat arbitrary, it should be a reasonably good approximation. For most firms in these industries, it is evident that $\tilde{R}_i(t)/S_i(t)$ increased during most of the postwar period and that it was a reasonably smooth function of time. Equation (13) implies that $\rho_i(t)$ was related in a particular way to t. It is difficult to say much about the "reasonableness" of this relationship.

Table 2.2. *Estimates of β's, Correlation Coefficients, and Percentage Error in Forecasting 1959 R and D Expenditures, Nineteen Firms, Chemical and Petroleum Industries*

	ESTIMATES[a]				STANDARD ERRORS				CORRE-LATION CO-EFFICIENT	FORECAST ERRORS (percent)
COMPANY	β_{i1}	β_{i2}	β_{i3}	β_{i4}	β_{i1}	β_{i2}	β_{i3}	β_{i4}		
Chemicals										
C1	0.601[b]	0.0068[b]	0.00138[b]	−0.00009[b]	0.198	0.0029	0.00059	0.00003	0.991	8.1
C2	.612[b]	.0197[b]	—	—	.171	.0074	—	—	.979	1.2
C3	.550[b]	.0091[b]	.00037[b]	—	.057	.0012	.00010	—	.999	−11.4
C4	.863[b]	.0022	—	.00007[b]	.179	.0028	—	.00003	.985	−9.3
C5	.811[b]	.0097[b]	—	—	.149	.0052	—	—	.968	0.8
C6	.644[b]	.0063[b]	.00122[b]	—	.155	.0029	.00049	—	.992	−3.1
C7	.768[b]	−.0015	.00300[b]	−.00018[b]	.243	.0052	.00132	.00008	.980	33.9
C8	.595[b]	.0286[b]	−.00092[b]	—	.201	.0114	.00040	—	.882	10.5
C9	−.476[b]	.0437[b]	−.00243[b]	.00031[b]	.221	.0075	.00108	.00008	.992	—[c]
C10	.496[b]	.0104[b]	.00095[b]	—	.109	.0024	.00029	—	.996	2.7
Total[d]	—	—	—	—	—	—	—	—	—	0.4
Petroleum										
P1	.684[b]	.0037[b]	−.00042[b]	.00003[b]	.146	.0014	.00020	.00001	.993	−6.3
P2	.727[b]	.0021[b]	—	—	.086	.0005	—	—	.994	3.6
P3	.875[b]	.0010	—	—	.164	.0007	—	—	.976	−7.2
P4	.044	.0044[b]	.00090[b]	—	.226	.0018	.00027	—	.965	−32.3
P5	.049	.00003	.00051[b]	—	.361	.0006	.00021	—	.993	−6.5
P6	.603[b]	.0002	.00073[b]	−.00004[b]	.203	.0017	.00036	.00002	.990	11.6
P7	.654[b]	.0025[b]	−.00024[b]	.00003[b]	.206	.0009	.00010	.00001	.996	−5.0
P8	.526[b]	.0023[b]	−.00016[b]	.000013[b]	.123	.0004	.00005	.000003	.998	−1.3
P9	.369[b]	.0044[b]	−.00021[b]	.000022[b]	.126	.0008	.00010	.000007	.998	−1.7
Total[d]	—	—	—	—	—	—	—	—	—	−4.1

SOURCE: See Appendix A.

[a] Statistically nonsignificant values of β_{i4} are omitted. Unless β_{i4} is significant, nonsignificant values of β_{i3} are omitted, too.

[b] Significantly different from zero (0.05 probability level).

[c] This firm is omitted because the estimate of β_{i1} is negative and significant. See Appendix A.

[d] The percentage error in forecasting the sum of these firms' 1959 R and D expenditures, using the sum of the forecasts for the individual firms.

few degrees of freedom, about three-fourths are statistically significant. For two-thirds of the firms, the estimates of β_{i3}, β_{i4}, or both are statistically significant as well.[21]

The second step is to use the results in Tables 2.2 and 2.3 to obtain estimates of $\tilde{R}_i(t)/S_i(t)$. Letting $\hat{\beta}_{i1}$, $\hat{\beta}_{i2}$, $\hat{\beta}_{i3}$, and $\hat{\beta}_{i4}$ be our estimates of β_{i1}, β_{i2}, β_{i3}, and β_{i4}, we use $\hat{\alpha}_{i1} = \hat{\beta}_{i2}/(1 - \hat{\beta}_{i1} + \hat{\nu}_4)$, $\hat{\alpha}_{i2} = \hat{\beta}_{i3}/(1 - \hat{\beta}_{i1} + \hat{\nu}_4)$, and $\hat{\alpha}_{i3} = \hat{\beta}_{i4}/(1 - \hat{\beta}_{i1} + \hat{\nu}_4)$ as estimates of α_{i1}, α_{i2}, and α_{i3}. We insert these estimates into equation (13) and ignoring $u_i(t)$, obtain estimates of $\tilde{R}_i(t)/S_i(t)$. These results are very rough, because the two principal assumptions are only approximations, $u_i(t)$ is ignored, the estimates of the α's are consistent but biased, and the presence of a lagged endogenous variable in equation (15)

[21] Judging by the correlation coefficients in Tables 2.2 and 2.3 equation (15) fits very well. However, if $R_i(t) - R_i(t-1)$ had been used instead of $R_i(t)$ as the dependent variable, these coefficients almost certainly would have been lower.

Table 2.3. *Estimates of β's, Correlation Coefficients, and Percentage Error in Forecasting 1959 R and D Expenditures, Sixteen Firms, Drug, Steel, and Glass Industries*

	ESTIMATES[a]				STANDARD ERRORS				CORRELATION COEFFICIENT	FORECAST ERRORS (percent)
COMPANY	β_{i1}	β_{i2}	β_{i3}	β_{i4}	β_{i1}	β_{i2}	β_{i3}	β_{i4}		
Drugs										
D1	0.866[b]	0.0099[b]	—	—	0.153	0.0051	—	—	0.969	11.0
D2	1.03[b]	.0045	—	—	.141	.0046	—	—	.977	−0.1
D3	.564[b]	.0448[b]	−.00579	.00034[b]	.302	.0232	.00303	.00016	.993	−2.2
D4	.469[b]	.0288[b]	—	—	.164	.0075	—	—	.960	2.8
D5	.490[b]	−.0057	.00444[b]	—	.175	.0074	.00110	—	.971	−6.3
D6	.366	.0386[b]	−.00275[b]	.00032[b]	.243	.0108	.00135	.00009	.999	−3.1
D7	.847[b]	.0222[b]	—	—	.112	.0081	—	—	.995	4.9
D8	.770[b]	.0297[b]	—	—	.114	.0064	—	—	.992	9.8
Total[c]	—	—	—	—	—	—	—	—	—	0.4
Steel										
S1	.257	.0018[b]	—	.000002[b]	.204	.0004	—	.000001	.978	−11.9
S2	.954[b]	.0000	.00017[b]	—	.089	.0008	.00009	—	.990	19.1
S3	.750[b]	−.0003	.00037[b]	—	.123	.0008	.00014	—	.980	19.3
S4	−.454	−.0004	.00014[b]	—	.394	.0002	.00004	—	.903	1.3
Total[c]	—	—	—	—	—	—	—	—	—	6.6
Glass										
G1	.993[b]	.0026	—	—	.132	.0022	—	—	.983	5.6
G2	.505[b]	.0127[b]	—	—	.162	.0034	—	—	.981	3.4
G3	.880[b]	.0043	—	—	.115	.0025	—	—	.947	−13.5
G4	−.030	.0230[b]	.00117[b]	—	.285	.0067	.00049	—	.947	−5.2
Total[c]	—	—	—	—	—	—	—	—	—	−1.7

SOURCE: See Appendix A.

[a] Statistically nonsignificant values of $\hat{\beta}_{i4}$ are omitted. Unless $\hat{\beta}_{i4}$ is significant, nonsignificant values of $\hat{\beta}_{i3}$ are omitted, too.

[b] Significantly different from zero (0.05 probability level).

[c] The percentage error in forecasting the sum of these firms' 1959 R and D expenditures, using the sum of the forecasts for the individual firms.

causes well-known statistical complications. Relevant details are presented in Appendix A.

The results contained in Table 2.4 indicate at least four things. First, the firms' desired R and D expenditures exceeded their actual R and D expenditures in almost all cases, the difference generally being greatest in the glass and drug industries and greater at the beginning than at the end of the period. Second, because of the differences between $\tilde{R}_i(t)$ and $R_i(t)$, actual R and D expenditures increased enormously during the period, the mean of $R_i(t)/S_i(t)$ increasing by 70 percent in chemicals, 50 percent in petroleum, 90 percent in drugs, 290 percent in steel, and 150 percent in glass.[22]

[22] Note three things: First, the increase in $R_i(t)/S_i(t)$ may be due partly to differences over time in the definition of R and D. Second, the 1954 changes in the tax treatment of R and D increased their profitability. Third, in comparing values of $\tilde{R}_i(t)/S_i(t)$ at various points in time, note that they are unadjusted for differences in the purchasing power of the dollar. For a discussion of the problems involved in making such adjustments, see H. Milton, "Cost-of-Research Index," *Operations Research* (November, 1966).

Third, at a given point in time, both $R_i(t)/S_i(t)$ and $\tilde{R}_i(t)/S_i(t)$ vary considerably among firms in the same industry, the standard deviation of $R_i(t)/S_i(t)$, which exceeds one-fourth of the mean in all but one of these industries, generally being smaller than the standard deviation of $\tilde{R}_i(t)/S_i(t)$. Fourth, in all industries but steel, the standard deviation of $R_i(t)/S_i(t)$ increased less rapidly during 1945–1958 than did the mean, the result being that the relative dispersion decreased during this period.[23]

In conclusion, Tables 2.2, 2.3, and 2.4 provide an exceptionally detailed and full description of the "research revolution" that occurred in these firms during 1945–1958. Although the results pertain only to thirty-five firms, they should be of considerable value to economists interested in the postwar behavior of R and D expenditures.[24]

4. Forecasts of Research and Development Expenditures

How accurate is equation (15) as a short-run forecasting device? To find out, we "forecasted" the 1959 R and D expenditures of each of the firms in Tables 2.2 and 2.3, using their 1959 sales, their 1958 R and D expenditures, and the estimates of β_{i1}, β_{i2}, β_{i3}, and β_{i4} in Tables 2.2 and 2.3. The results, shown in Tables 2.2 and 2.3, indicate that the forecasting errors for individual firms were generally less than 10 percent, that they were only about 3 percent for the industry totals, and that they were considerably smaller than those resulting from three standard naïve models. (See Appendix A.) Of course, this comparison is somewhat unfair because it assumes that we have perfect information regarding the firm's 1959 sales, but when forecasts were based on 1959 sales data containing 10 percent errors, the results were still superior to the naïve models.

Given these encouraging results, we used equation (15) to make some longer-run forecasts, supposing this time that we were forecasting in 1959 on the basis of two sets of naïve assumptions. First, we assumed that α_{i1}, α_{i2}, α_{i3}, ν_3, ν_4, ν_5, and π_i would remain at their 1945–1958 levels during the 1960's—which means in effect that $\tilde{R}_i(t)/S_i(t)$ would continue to conform to equation (13). Second, we assumed that ν_3, ν_4, ν_5, and π_i would remain at their 1945–1958 levels, but that $\tilde{R}_i(t)/S_i(t)$ would stay at its 1959 level. Under both sets of assumptions, we supposed that for 1960–1969,

$$S_i(t) = S_i(1 + r)^{t'}, \tag{16}$$

where S_i is the ith firm's sales in 1959; t' is measured in years from 1959; and $r = 0.04$ for petroleum firms, 0.05 for chemical firms, and 0.06 for drug firms.[25]

[23] The customary measure of relative dispersion is the standard deviation divided by the mean.

[24] Note that these indirect estimates of $\tilde{R}_i(t)/S_i(t)$ agree quite well with those in Table 2.1. (See Appendix A.) This is an important test of these estimates.

[25] Using equation (16), I generate 1960–1969 figures for $S(t)$. To obtain the first set of forecasts, I insert these figures and the $\hat{\beta}$'s into equation (15). To obtain the second set, I do the same thing except that t is fixed at 15.

Table 2.4. *Mean and Standard Deviation of $R_i(t)/S_i(t)$ and $\tilde{R}_i(t)/S_i(t)$, Chemical, Petroleum, Drug, Steel, and Glass Firms, 1945–1958*

INDUSTRY	1945	1946	1947	1948	1949	1950
Chemical firms (9)						
Mean of $R_i(t)/S_i(t)$	0.0238	0.0302	0.0286	0.0276	0.0308	0.0265
Mean of $\tilde{R}_i(t)/S_i(t)$	.02687	.02798	.02962	.03116	.03258	.03391
S.D. of $R_i(t)/S_i(t)$	.0092	.0116	.0131	.0121	.0136	.0126
S.D. of $\tilde{R}_i(t)/S_i(t)$	.01971	.01755	.01486	.01253	.01058	.00891
Petroleum firms (9)						
Mean of $R_i(t)/S_i(t)$	.00558	.00658	.00581	.00510	.00596	.00547
Mean of $\tilde{R}_i(t)/S_i(t)$	.00525	.00533	.00545	.00560	.00578	.00599
S.D. of $R_i(t)/S_i(t)$	.00240	.00278	.00158	.00098	.00142	.00159
S.D. of $\tilde{R}_i(t)/S_i(t)$	.00315	.00255	.00210	.00184	.00175	.00179
Drug firms (7)						
Mean of $R_i(t)/S_i(t)$	.0358	.0351	.0460	.0494	.0511	.0538
Mean of $\tilde{R}_i(t)/S_i(t)$	.07190	.06970	.06869	.06836	.06837	.06874
S.D. of $R_i(t)/S_i(t)$	.0158	.0120	.0163	.0150	.0149	.0161
S.D. of $\tilde{R}_i(t)/S_i(t)$	.03935	.03871	.03700	.03516	.03372	.03253
Steel firms (3)						
Mean of $R_i(t)/S_i(t)$	.00302	.00328	.00285	.00323	.00398	.00629
Mean of $\tilde{R}_i(t)/S_i(t)$	.00233	.00264	.00334	.00406	.00478	.00550
S.D. of $R_i(t)/S_i(t)$	.00096	.00102	.00006	.00064	.00112	.00436
S.D. of $\tilde{R}_i(t)/S_i(t)$	.00330	.00309	.00262	.00217	.00183	.00163
Glass firms (3)						
Mean of $R_i(t)/S_i(t)$	.0111	.0130	.0157	.0180	.0200	.0186
Mean of $\tilde{R}_i(t)/S_i(t)$	.0418	.0418	.0418	.0418	.0418	.0418
S.D. of $R_i(t)/S_i(t)$	.0045	.0064	.0041	.0045	.0026	.0048
S.D. of $\tilde{R}_i(t)/S_i(t)$	.0216	.0216	.0216	.0216	.0216	.0216

SOURCE: See section 3.

NOTE: The estimates of $\tilde{R}_i(t)/S_i(t)$ are indirect estimates based on the assumptions described in section 3. A few firms in Tables 2.2 and 2.3 had to be omitted because $[1 - \hat{\beta}_{i1} + \hat{\nu}_4] < 0$.

The results depend heavily on which set of assumptions one makes. If $\tilde{R}_i(t)/S_i(t)$ continued to conform to equation (13) during the 1960's, Table 2.4 shows that the ratio of R and D expenditures to sales would increase even more rapidly than during the 1950's. However, if $\tilde{R}_i(t)/S_i(t)$ remained at its 1959 level, only rather modest increases could be expected in the ratio of R

1951	1952	1953	1954	1955	1956	1957	1958
0.0277	0.0336	0.0341	0.0366	0.0332	0.0350	0.0373	0.0403
.03512	.03623	.03723	.03812	.03891	.03959	.04017	.04063
.0135	.0117	.0103	.0115	.0091	.0077	.0081	.0099
.00768	.00702	.00720	.00838	.01040	.01317	.01654	.02050
.00538	.00611	.00631	.00652	.00686	.00687	.00785	.00826
.00623	.00650	.00680	.00714	.00750	.00790	.00832	.00878
.00197	.00252	.00234	.00223	.00238	.00278	.00413	.00343
.00192	.00207	.00223	.00237	.00255	.00274	.00300	.00331
.0540	.0586	.0559	.0633	.0592	.0588	.0624	.0697
.06944	.07050	.07189	.07364	.07573	.07816	.08095	.08407
.0203	.0182	.0177	.0219	.0126	.0154	.0189	.0188
.03156	.03067	.02986	.02902	.02821	.02744	.02674	.02629
.00479	.00528	.00487	.00706	.00750	.00742	.00847	.01172
.00623	.00696	.00770	.00845	.00920	.00996	.01072	.01148
.00246	.00227	.00217	.00345	.00395	.00317	.00402	.00636
.00161	.00178	.00207	.00244	.00287	.00331	.00383	.00428
.0185	.0189	.0197	.0240	.0226	.0247	.0255	.0273
.0418	.0918	.0418	.0418	.0418	.0418	.0418	.0418
.0037	.0036	.0012	.0023	.0047	.0039	.0044	.0010
.0216	.0216	.0216	.0216	.0216	.0216	.0216	.0216

and D expenditures to sales.[26] (See Table 2.5.) From the National Science

[26] The results in sections 3 and 4 assume that the firms continually operate in the range where equation (9) holds. On the basis of the estimates in equation (11), the value of $\pi_i(t-1)$ observed during 1945–1958 in the chemical industry, and what little we can guess about $(\tilde{R}_i(t) - R_i(t-1))/R_i(t-1)$, it seems likely that this was the case for most firms. But for the remaining minority, it is possible that this equation did not hold and that $\theta_i(t)$ was usually one. For such firms, the estimates of $\tilde{R}_i(t)/S_i(t)$ will be underestimated slightly but the results in this section will not be affected so long as $\theta_i(t)$ continues to be one.

For another set of forecasts, see D. Keezer, D. Greenwald, and R. Ulin, "The Outlook for Expenditures on Research and Development During the Next Decade," *American Economic Review*, May 1960. A few years ago, statements by government and business officials seemed to imply a decrease in $\tilde{R}_i(t)/S_i(t)$, at least in some industries. If true in these industries, even the second set of forecasts is likely to be too high.

Table 2.5. *R and D Expenditures as a Percent of Sales, Chemical, Petroleum, and Drug Firms, Actual (1950–1959) and Projected (1960–1969), Under Two Sets of Assumptions*

	ACTUAL R AND D EXPENDITURES									
INDUSTRY	1950	1951	1952	1953	1954	1955	1956	1957	1958	1959
Chemicals	2.73	2.69	3.35	3.45	3.90	3.50	3.73	4.12	4.36	4.14
Petroleum	0.55	0.54	0.59	0.62	0.59	0.60	0.60	0.66	0.73	0.74
Drugs	4.53	4.27	5.15	5.20	5.47	5.51	5.61	5.96	6.72	7.11
	FORECASTS OF R AND D EXPENDITURES (ASSUMPTION I)									
INDUSTRY	1960	1961	1962	1963	1964	1965	1966	1967	1968	1969
Chemicals	4.61	4.66	4.80	4.90	5.04	5.18	5.35	5.57	5.79	6.02
Petroleum	0.83	0.88	0.93	0.98	1.05	1.12	1.23	1.34	1.45	1.58
Drugs	7.57	8.23	8.87	9.53	10.22	10.94	11.71	12.52	13.38	14.27
	FORECASTS OF R AND D EXPENDITURES (ASSUMPTION II)									
INDUSTRY	1960	1961	1962	1963	1964	1965	1966	1967	1968	1969
Chemicals	4.49	4.51	4.52	4.53	4.53	4.54	4.54	4.55	4.55	4.55
Petroleum	0.82	0.84	0.85	0.85	0.86	0.86	0.87	0.87	0.87	0.87
Drugs	7.32	7.59	7.76	7.88	7.98	8.05	8.12	8.18	8.23	8.28

SOURCE: See section 4.

NOTE: $\sum_i R_i(t)/\sum_i S_i(t)$, not the unweighted average value of $R_i(t)/S_i(t)$, is shown here.

Foundation's data for 1960–1963, it would appear that the latter forecasts are more realistic.

5. Size of Firm and Expenditures on Research and Development

It is often maintained that a policy designed to break up large firms would be a mistake because bigness is required to carry on the research and development which is vital to rapid technological change. No one would deny that some minimum size is needed before a firm can maintain a profitable and effective R and D program, but are giants like U.S. Steel, Standard Oil (N.J.), and DuPont required for this purpose? The remainder of this chapter is concerned with three very important questions that bear on this issue. (1) Do the largest firms spend more on R and D, relative to their size, than firms of one-fifth or one-tenth of their size? (2) Within the relevant range, do increases in R and D expenditures result in more than proportional increases in "inventive output"? (3) When the level of R and D expenditures is held constant, is "inventive output" higher in the largest firms than in somewhat smaller ones?

This section, which deals with the first question, investigates the relationship between the size of the firm and the level of R and D expenditures among the major firms in the chemical, petroleum, drug, steel, and glass industries during 1945–1959. In each industry, assume that

$$ln\ R_i(t) = \phi_o(t) + (\phi_1 + \phi_2 t)\ ln\ S_i(t) + z_i'''(t), \tag{17}$$

where $z_i'''(t)$ is an error term and time is measured from 1945. This equation is used simply as a convenient description of the data, which pertain to ten chemical firms, nine petroleum firms, eight drug firms, seven steel firms, and four glass firms. The important parameters are ϕ_1, which measures the 1945 elasticity of R and D expenditures with respect to size of firm, and ϕ_2, which measures the annual change in the elasticity during 1945–1959.

Least-squares estimates of ϕ_1, shown in Table 2.6, indicate that the largest

Table 2.6. *Estimates of $\phi_o(t)$ and ϕ_1, Chemical, Petroleum, Drug, Steel, and Glass Industries, 1945–1959*

PARAMETER	CHEMICALS (10 FIRMS)	PETROLEUM (9 FIRMS)	DRUGS (8 FIRMS)	STEEL[a] (7 FIRMS)	GLASS (4 FIRMS)
ϕ_1	1.10 (.02)	0.86 (.04)	0.91 (.04)	0.90 (.14)	0.87 (.06)
ϕ_o (1959)	−3.83	−3.86	−2.23	—	−2.84
ϕ_o (1958)	−3.75	−3.85	−2.30	—	−2.77
ϕ_o (1957)	−3.84	−3.92	−2.43	−4.52	−2.86
ϕ_o (1956)	−3.93	−4.04	−2.48	—	−2.96
ϕ_o (1955)	−3.99	−4.05	−2.49	−4.94	−2.98
ϕ_o (1954)	−3.92	−4.10	−2.47	—	−2.97
ϕ_o (1953)	−4.00	−4.15	−2.57	—	−3.07
ϕ_o (1952)	−4.02	−4.21	−2.56	—	−3.20
ϕ_o (1951)	−4.21	−4.33	−2.70	—	−3.18
ϕ^o (1950)	−4.27	−4.34	−2.70	−5.33	−3.25
ϕ_o (1949)	−4.05	−4.22	−2.75	—	−3.20
ϕ_o (1948)	−4.20	−4.38	−2.81	—	−3.32
ϕ_o (1947)	−4.13	−4.28	−2.94	—	−3.42
ϕ_o (1946)	−4.00	−4.21	−3.18	−5.69	−3.57
ϕ_o (1945)	−4.24	−4.43	−3.16	—	−3.72
r	0.98	0.93	0.93	0.82	0.92

SOURCE: See section 5.

NOTE: The results pertain only to large firms in these industries. The standard error of ϕ_1 is shown in parentheses.

[a] The steel data, which come from A. Ninian, pertain to only 4 years. See his "The Role of Research in the American Steel Industry" (unpublished manuscript, M.I.T. thesis, 1959). These data were used rather than those underlying Tables 2.3 and 2.4 because results could be obtained for a larger number of firms.

firms in most of these industries spent less in 1945 on R and D, relative to sales, than did somewhat smaller firms. In petroleum, drugs, and glass, the largest firms spent less (ϕ_1 being significantly less than one); in chemicals, they spent more (ϕ_1 being significantly greater than one); and in steel, they spent less but the difference is not statistically significant.

There is no evidence that the effect of size of firm on the level of R and D expenditures changed systematically during 1945–1959, the estimate of ϕ_2 being statistically nonsignificant in every industry. This is interesting because one might suppose, like Schmookler [144], that smaller firms would increase their R and D expenditures more rapidly than the largest ones, the result being that $\phi_2 < 0$. There is no evidence that this occurred.[27]

In summary, except for the chemical industry, the largest firms in these industries seemed to spend no more on R and D, relative to sales, than did somewhat smaller firms. During 1945–1959, there is no evidence that this situation changed appreciably.[28]

6. Research and Development Expenditures, Size of Firm, and Inventive Output

What is the relationship between the level of a firm's R and D expenditures and the number of significant inventions it produces? Holding the level of R and D expenditures constant, is "inventive output" higher in the largest firms than in somewhat smaller ones? The answers to these questions are extremely important because they indicate the extent to which there are economies of scale in R and D[29] and the extent to which the largest firms are more efficient than others in conducting an R and D program of given scale. This section presents some very tentative findings regarding the chemical, petroleum, and steel industries.

In each industry, assume that

$$n_i = R_i[a + bR_i + cS_i] + V_i, \tag{18}$$

where n_i is the weighted number of inventions, or innovations, carried out by the ith firm, R_i is its R and D expenditures during the relevant period, S_i

[27] Another hypothesis might be that ϕ_2 is negative if $\phi_1 > 1$ and positive if $\phi_1 < 1$. There is no evidence that this hypothesis is true either.

[28] Since this work was completed, a similar study has been carried out by Scherer, the results being like those presented above. See F. Scherer, Testimony before the Senate Subcommittee on Antitrust and Monopoly, May 25, 1965, and "Firm Size, Market Structure, Opportunity and the Output of Patented Inventions," *American Economic Review* (December, 1965). Of course, the largest firms probably do a somewhat different kind of R and D than smaller ones. All R and D expenditures are lumped together here. See Chap. 3.

[29] Economies of scale may be present because of the lumpiness of equipment, the advantages of using highly specialized personnel, and the greater chance that successes and failures will cancel out. See Chap. 5 and E. Mansfield, *Monopoly Power and Economic Performance* (New York: Norton, 1964).

is its size (measured in terms of sales), and V_i is a random error term.[30] In the chemical industry, n_i is measured by Langenhagen's data [72] regarding the number of significant inventions (weighted roughly by a measure of their importance) carried out by each firm between 1940 and 1957. Results based on data regarding ten major chemical firms are

$$\text{(19)} \qquad n_i = R_i[\underset{(0.041)}{2.38} + \underset{(0.075)}{0.404}R_i - \underset{(0.0055)}{0.0247}S_i], \quad (r = 0.99)$$

where R_i is the average of the ith firm's expenditures on R and D in 1940 and 1950, S_i is its sales in 1940, and V_i is omitted.

In the petroleum industry, we use Schmookler's list [145] of important inventions in petroleum refining and my list of important petrochemical innovations (Table 5.2) to measure n_i. Data for eight major firms indicate that

$$\text{(20)} \qquad n_i = \underset{(0.027)}{0.508}R_i, \quad (r = 0.99)$$

where R_i is the average of the ith firm's expenditures on R and D in 1945 and 1950, n_i is the combined number of refining inventions and petrochemical innovations (weighted roughly by their importance) carried out by the ith firm between 1946 and 1956, b and c are set equal to zero because their estimates are not statistically significant, and V_i is omitted.

In the steel industry, my list of important innovations (Table 5.1) is used, n_i being the number of these innovations carried out by the ith firm between 1946 and 1958. Data for eleven major firms indicate that

$$\text{(21)} \qquad n_i = R_i[\underset{(0.27)}{1.19} - \underset{(0.000212)}{0.000548}S_i], \quad (r = 0.87)$$

where R_i is the average of the ith firm's expenditures on R and D in 1946 and 1950, S_i is its sales in 1946, b is set equal to zero because its estimate is not statistically significant, and V_i is omitted.

Equations (19) to (21) suggest at least three things. First, when the size of the firm is held constant, the number of significant inventions carried out by

[30] In effect, we assume that the expected number of inventions per R and D dollar is a linear function of R_i and S_i. But since there is no evidence that V_i is heteroscedastic, we use equation (18), rather than an expression for n_i/R_i, to estimate a, b, and c. However, if n_i/R_i had been used, the results would have been much the same, although the correlation coefficients would have been lower.

The data underlying equations (19) to (21) are described in Appendix A. R_i and S_i are measured in units of millions of dollars. Note that there may be biases in the estimates of a, b, and c because R_i is not entirely exogenous and least-squares estimates are used. Moreover, the "productivity" of a firm's R and D expenditures obviously depends on the sort of R and D it carries out. For some petroleum firms, research on drilling and production really should be excluded.

a firm seems to be strongly influenced by the size of its R and D expenditures. The partial correlation of n_i on R_i, holding S_i constant, is 0.96, 0.98, and 0.70 in chemicals, petroleum, and steel. Thus, although the pay-out from an individual R and D project is obviously very uncertain, there seems to be a close relationship over the long run between the amount a firm spends on R and D and the total number of important inventions it produces.

Second, when R and D expenditures are held constant, the effects of firm size on the average productivity of such expenditures turn out to be negative in each industry and statistically significant in two of the three industries. Thus, contrary to popular belief, the inventive output per dollar of R and D expenditure in most of these cases seems to be lower in the largest firms than in large and medium-sized firms. In part, this may be due to looser controls and greater problems of supervision and coordination in a very large organization.[31]

Third, when the size of the firm is held constant, the evidence seems to suggest that increases in R and D expenditures result in more than proportional increases in inventive output in the chemical industry. However, in petroleum and steel, there is no real indication that b is positive. Thus, except for chemicals, the results do not indicate any marked advantages of the largest-scale research activities over large and medium-sized ones.[32]

Finally, the crudeness of these results should be noted. The measures of "inventive output" are obviously only the roughest approximations, since they include only inventions that are considered significant in some sense by informed observers and these inventions are weighted arbitrarily. Moreover, because of differences in accounting procedures and errors of measurement, the data regarding R and D expenditures may not be entirely comparable from one firm to another.[33]

7. Summary and Conclusions

The principal conclusions of this chapter are as follows: First, based on the small amount of data that could be obtained, there is some evidence that the level of a firm's research and development expenditures can be explained reasonably well by a simple model which assumes that (a) the expected rates of return from reasonably promising research and development projects are

[31] For some relevant discussion in this connection, see Mansfield, *Economics of Technological Change*, Chap. 3; and Jewkes, Sawers, and Stillerman, *Sources of Invention*, Chaps. 5 to 8.

[32] Most of the firms that are included here spent a reasonably large amount on R and D, generally several hundred thousands of dollars a year or more. There may be considerable economies of scale in the lower ranges of spending.

[33] For some discussion of the problems involved in measuring inventive input and output, see S. Kuznets, "Inventive Activity: Problems of Definition and Measurement," *The Rate and Direction of Inventive Activity* (Princeton, 1962); and F. Machlup, "The Supply of Inventors and Inventions," *The Rate and Direction of Inventive Activity*, *ibid.* Note too that the weights used in equations (19) and (20) are based on the economic importance of various inventions to the whole industry, rather than to the inventor. For some purposes, it might have been useful to have based them on the profitability to the inventor too.

distributed according to the Pareto law, (b) the distribution of expected rates of return from research and development projects, together with a firm's size, determines the firm's desired level of research and development expenditures, and (c) the firm's speed of response toward this desired level depends on the extent to which the desired level differs from last year's spending and the percent of its profits spent last year on research and development.

Second, this model may be of use in forecasting research and development expenditures and in estimating the effects on research and development expenditures of changes in relevant public policies. For example, there is some indication that a change in policy increasing the profitability of research and development projects by 1 percent would have resulted in recent years in a 2 percent increase in the research and development expenditures of large petroleum firms and a $3\frac{1}{2}$ percent increase in the research and development expenditures of large chemical firms. (These increases are in money terms.) However, it must be emphasized that the data are so limited in quality and quantity that these results should be taken with a grain of salt.

Third, if, because of lack of data regarding $\bar{\rho}_i(t)$, one assumes that a firm's desired research and development expenditures, as a percent of sales, are a quadratic function of time, the resulting version of the model seems quite useful for short-term forecasting. Results for thirty-five firms in the chemical, petroleum, drug, steel, and glass industries indicate that the model does considerably better than the customary naïve models. The model is also used experimentally to forecast research and development expenditures, as a percent of sales, in several industries during the late 1960's, the results depending heavily on which of two sets of assumptions is used.

Fourth, except for the chemical industry, there is no evidence that the largest firms in these industries spent more on research and development, relative to sales, than did somewhat smaller firms. In the petroleum, drug, and glass industries, the largest firms spent significantly less; in the steel industry, they spent less but the difference was not statistically significant. During 1945–1959, there was no indication in any of these industries of a systematic change in the effect of size of firm on the level of research and development expenditures.

Fifth, where the size of the firm is held constant, the number of significant inventions carried out by a firm seems to be highly correlated with the level of its research and development expenditures. In the chemical industry, increases in research and development expenditures apparently result in more than proportional increases in inventive output, but in petroleum and steel, there is no evidence of either economies or diseconomies of scale within the relevant range. In most industries, the productivity of a research and development program of given scale seems to be lower in the largest firms than in somewhat smaller firms.

CHAPTER 3

Allocation, Characteristics, and Outcome of the Firm's Research and Development Portfolio: A Case Study

The previous chapter explored the factors influencing the size of a firm's research and development expenditures and the relation of such expenditures to the firm's inventive output. However, little was said about the allocation of funds among research and development projects, the characteristics of the projects that are undertaken, and the probable outcome of these projects.[1] The purpose of this chapter is to discuss these subjects, which are both important and relatively unexplored. Despite the remarkable increase in recent years in the amount of attention devoted to the economics of research and development, there have been surprisingly few detailed studies of the research and development activities of the firm.

More specifically, this chapter reports the findings of a case study of the research and development portfolio of the central research laboratory of one of the nation's largest firms, a prominent electrical and electronic equipment and appliance manufacturer. The study, which took 2 years, is the most detailed treatment to date of the allocation, characteristics, and outcome of a firm's research and development expenditures. Data were obtained regarding seventy major projects, and numerous interviews were obtained with officials at various levels of the firm. Like all case studies, the results are limited by

This chapter is based on work done jointly with Richard Brandenburg.

[1] Although there is a growing literature proposing techniques for R and D project selection, it appears that such procedures have been difficult to implement. For example, see N. Baker and W. Pound, "R and D Project Selection: Where We Stand," presented at the joint TIMS-ORSA Meeting, Minneapolis, Oct. 1964. Further, little descriptive work has been done. See R. Brandenburg, "Research and Development Project Selection" (Ph.D thesis, Cornell, 1964), for a first step along these lines.

the fact that they pertain only to one firm, but where possible the findings were checked against preliminary results for other firms.[2]

In section 1, we describe the process by which research and development proposals and budgets were generated and evaluated within this laboratory. In section 2, we present and test a simple model to explain the expenditures proposed for a particular project; section 3 contains a similar model to explain the modifications made by the laboratory management in the proposed level of expenditures; the implications of the results are discussed in section 4. Section 5 describes the characteristics of the projects that were accepted; and in sections 6 and 7, we analyze the outcomes of these projects and the extent to which the firm was able to predict their technical success or failure.

1. The Decision Process Within the Firm

The central research laboratory of the firm we studied is organized into departments concerned with the study of materials and equipment, other aspects of relevant technology, and basic science.[3] The bulk of the laboratory's expenditures, totaling about $20 million annually, are for applied R and D, only about 10 percent being for basic research. Excluding basic research, about 60 percent of the laboratory's expenditures are on "new business" projects which attempt to develop entirely new products and almost all of which stem from the laboratory. The remaining 40 percent of the expenditures (excluding basic research) are on projects suggested by operating divisions of the firm, which request particular product improvements and new processes.

This section describes how the laboratory's portfolio of applied R and D projects was chosen in 1963,[4] the first step in this process being a rough screening by the laboratory management of various project proposals. With regard to projects proposed by the divisions, this screening began in the summer of 1962, when the laboratory asked the divisions for proposals for

[2] I am carrying out a study of the characteristics of the R and D portfolios of chemical and petroleum firms. Although the results are not yet ready for publication, the work is far enough along to provide some indication of the representativeness of some of the findings presented below. A. Rubenstein, in a paper presented at the Institute of Management Science in June 1959 ("Rate of Organizational Change, Corporate Decentralization, and the Constraints on Research and Development in the Firm"), and in collaboration with R. Hannenberg ("Idea Flow and Project Selection in Several Industrial Research and Development Laboratories," *The Economics of Research and Development*, Ohio State, 1965), has also presented some relevant material.

[3] Examples of these departments are "materials science," "metallurgy and ceramics technology," "science and mathematics," and "equipment science." For a discussion of the organization and administration of R and D in various firms, and the problems involved, see E. Mansfield, *The Economics of Technological Change* (New York: Norton, 1968), Chap. 3.

[4] Throughout the section, we are concerned for obvious reasons with applied R and D, not basic research. The section contains data regarding 1964 as well as 1963. The procedure for selecting projects was somewhat different in 1964.

1963. For each research proposal, a division was requested to estimate (1) the probability of commercial success of the project (if technically successful), (2) the extra profit to the firm if the project was commercially successful, and (3) the investment required to put the research results into practice.[5] These proposals were then sent to the managers of the relevant laboratory departments, who made preliminary estimates of the cost of doing the R and D and the corresponding chance of technical success. Combining the information provided by the division and the department manager, the laboratory's project evaluation group, a small group of project analysts that report to the laboratory management, computed a "figure of merit" for each proposal and rated it "A," "B," or "C." [6] The project evaluation group also computed a "figure of merit" for each "new business" project, using estimates of potential market and production costs for the product which the project would try to develop, and rated it as "A," "B," or "C."

The second step in the decision-making process involved the formulation of a proposed R and D portfolio by each department of the laboratory. The department manager received all division requests and new business proposals that fell within his department's responsibility, together with the rating of each project. The laboratory management recommended that the "A" proposals be given top priority, and that the "C" proposals be avoided, but beyond such loose guidelines the department manager was free to formulate whatever set of projects seemed best to him. After considerable deliberation, each department manager sent the laboratory management a list of projects that he wanted to carry out. These projects were designed to be largely independent entities.[7] For each project, the department manager proposed

[5] Note four things: First, in rating projects, the firm seemed to act as if there were only two outcomes of the project—"technical success" or "technical failure." If it was a "technical success," it was again assumed that there would be only two outcomes—"commercial success" or "commercial failure." A "technical failure" produces no useful information; a "commercial failure" produces no profits. Needless to say the model of the world implicit in this rating scheme was exceedingly simple—as the relevant executives realized. Second, the profit figures were gross of the costs involved in putting the research results into practice, i.e., the investment required. Third, other questions were also asked the division, but the three mentioned in the text seemed to be the most important. Fourth, there is sometimes a tendency for divisions to exaggerate the profitability of proposed projects that are of interest to them. Realizing this, the firm has adopted a number of procedures to check this tendency. In particular, a series of echelons within the divisions are asked to inspect the profitability estimates, and the project evaluation group within the laboratory makes an independent estimate of the profitability of each proposed project.

[6] For each project, the "figure of merit" equaled $qtrG/(E + D)$, where q is the project evaluation group's estimate of the extra annual profit if the project is technically and commercially successful, t is their estimate of the number of years that this profit stream will continue, G is their estimate of the probability of commercial success, r is the department manager's preliminary estimate of the probability of technical success, E is his preliminary estimate of the cost of the R and D, and D is the division's estimate of the cost of putting the research result into practice. If $M > 4$, the proposal was rated "A"; if $4 \geq M \geq 2$, the project was rated "B"; and if $M < 2$, the project was rated "C."

[7] That is, to the extent possible, the projects are defined so as to reduce their interdependence. For example, parallel paths to the same objectives are considered parts of the same project, not separate projects. Nonetheless, some interdependence remains and is recognized by the project evaluation group and the laboratory management.

a level of expenditure and estimated the probability of technical success. In the next section, this proposed level of expenditure is called C^* and the estimated probability of success is called P^*.

The final step in the decision-making process involved the modification of each department's proposals by the laboratory and corporate management. The laboratory management evaluated each project proposed by the department manager, decided whether to accept or reject it, and suggested a level of expenditures on each project that it accepted. (This level of expenditures is referred to as C^*_M in section 3.) A list of the projects that survived this stage of the process was submitted to the corporate management, together with the proposed expenditures on each one. The corporate management then set the level of total expenditures at the laboratory. Finally, since the level set by corporate management was too small to finance all of the projects on the list, some were dropped and others had their budgets modified.

What determined how much money was allocated to a particular R and D project? To help sort out the relevant factors, it is convenient to break this question into two parts: first, what determined the amount that the department manager proposed to spend on the project? Second, what determined the extent to which the laboratory management modified the department manager's proposal? Section 2, which is concerned with the first question, constructs and tests a simple model to explain the department manager's proposed level of expenditures. Section 3 takes up the second question.

At the outset, two points should be noted. First, the simple models presented in sections 2 and 3 are not meant to be descriptions, in any literal sense, of the decision-making process. In fact, of course, the relevant managers do not go through the process visualized by the models, but this is not important, so long as they act as if they do. Second, if the models are to be testable, they must be formulated in terms of concepts which correspond to those for which the firm generates data. In the previous section, we saw that the principal pieces of information generated within the firm are estimates of a project's expected profitability and its probability of success. It seems natural to build our models around these concepts, both because they are important and because data are available for them.

2. The Proposed Level of Expenditures on a Research and Development Project

We begin by assuming that the department manager believes that the probability that a project is successfully completed between time t and time $t + \Delta$ is $\beta C \Delta$, where C is the rate of expenditure on the project, β is a parameter that varies from project to project, and t is measured in years. If this extremely simple model holds,[8] it can be shown that

[8] Taking a particular project, it seems obvious that the probability of completing the project in a given time interval is an increasing function of the length of the time interval and the annual rate of expenditure on the project. Moreover, it is also obvious that the

(1) $$P = 1 - e^{-\beta C},$$

where P is the probability that the project will be successfully completed during the year. Thus, the expected discounted profit from the project is

(2) $$\pi = (1 - e^{-\beta C})X - C,$$

where X is the estimated discounted profit if the project is technically successful.[9]

Next assume that the department manager has in mind some total budget for his department and that he fixes C so as to maximize the sum of the expected discounted profits from all his projects, subject to this constraint. He assumes that the projects are independent and he sets

(3) $$\frac{\partial \pi}{\partial C} = \beta e^{-\beta C}X - 1 = \lambda,$$

where λ is the marginal expected discounted profit from an extra dollar spent in his department. Thus, the amount of money that he proposes to spend on the project is

(4) $$C^* = ln\left[\frac{\beta X}{1 + \lambda}\right] / \beta.$$

To test this model, we make two additional assumptions. First, we assume that the department manager's estimate of X is determined by X_D, the division's estimate of the discounted profits (net of division costs) if the project is successful.[10] In particular, suppose that

(5) $$X = \alpha_o X_D^{\alpha_1} Z,$$

---

probability tends to zero as either of these variables tends to zero. Given these conditions, the simplest assumption that meets them is that the probability is proportional to the product of the two variables. Obviously, the specific functional form that is used is mainly for analytical convenience. The heart of the model is the *general* hypothesis that as the rate of spending goes up, the chances of success go up too. This model is a reasonable approximation for R and D where the probability of success, holding the time interval and the rate of expenditures constant, does not vary greatly with time. In effect, it views R and D as a search process where the ground previously explored is always so small, relative to the total, that the probability of success varies little within a reasonably short span of time. Although not all R and D is of this type, this model is perhaps as good a first approximation as any of comparable simplicity.

[9] According to the model, the department manager assumes that if a project receives a budget of C for the coming year, it will spend it all—even if success is reached before the end of the year. As a first approximation, this probably does relatively little violence to reality. Also, we follow the practice, implicit in the firm's rating scheme, of assuming that a project which is not a "technical success" results in no profits, now or in the future.

[10] We restrict our attention to projects proposed by an operating division. The reader will recall from section 1 that the division was required to provide information regarding the probability of commercial success, the extra profit if the project was commercially successful, and the investment required. Using an interest rate of 0.10, we obtained X_D by discounting back the expected profits if the project was technically successful.

where the α's are the same for all projects and Z is a random error term. Second, we assume that λ, a nonobservable variable which undoubtedly differs from department manager to department manager, is statistically independent of X_D.

Given these assumptions, the model predicts that

$$\text{(6)} \qquad ln\left(\frac{C^*}{1-P^*}\right) - ln\left\{ln\left(\frac{1}{1-P^*}\right)\right\} = \phi + \alpha_1 \, ln \, X_D + Z^1,$$

where ϕ equals $ln \, \alpha_0$ plus the average value of $ln \, [1/(1+\lambda)]$, P^* is the probability of technical success associated with the level of expenditure (C^*) proposed by the department manager, and Z^1 is an error term which equals Z plus the deviation of $ln \, [1/(1+\lambda)]$ from its average value. To obtain equation (6), note that $\beta = \frac{ln \, [1/(1-P^*)]}{C^*}$, substitute this expression for β into equation (4), and then substitute the expression for X in equation (5) into the result.

To test the model, we gathered data regarding C^*, P^*, and X_D for eleven major projects;[11] and using equation (6), we obtained least-squares estimates of ϕ and α_1. The resulting regression,

$$\text{(7)} \qquad ln\left(\frac{C^*}{1-P^*}\right) - ln\left\{ln\left(\frac{1}{1-P^*}\right)\right\} = \underset{(0.44)}{4.13} + \underset{(0.06)}{0.21} \, ln \, X_D,$$

is quite consistent with the model, the regression coefficient having the predicted sign and being statistically significant. About 50 percent of the variation in the dependent variable can be explained. Thus, the results, though extremely tentative, seem to support the model reasonably well.[12]

3. Modifications by the Laboratory Management

To determine the extent to which the laboratory management modified the proposals made by the department manager, data were obtained regarding

[11] The sample size was reduced by the necessity to omit projects with some government financing (since the model obviously does not apply to them) and by the expense of gathering the detailed data underlying the estimates of X_D. See note 28 for other considerations. Although the number of projects is small, they account for quite a large dollar expenditure.

[12] Note five things: First, for given values of X_D, equation (7) provides an expected relationship between C^* and P^*; to predict C^*, one must know P^*. Of course, this does not lessen the usefulness of equation (7) as a test of the model. Second, there may be an identification problem in equation (7), since the department managers, in dealing with divisions, sometimes are able to influence x_D so as to rationalize certain values of C^*. Third, we eliminated β from equations (6) and (7) because, whereas C^* and P^* are variables actually measured and used by the firm, β does not exist outside of the model. It stems entirely from the assumption in equation (1). Fourth, interviews with laboratory executives indicate that they agreed with many of the basic ideas underlying the model, although not necessarily with the details. Fifth, all tests in this paper use the 0.05 significance level.

the disposition of seventy-two projects during 1963 and 1964.[13] The results, shown in Table 3.1, indicate that the modifications were generally substantial.

Table 3.1. *Changes by the Laboratory Management in the Proposed Budgets of Seventy-two R and D Projects*

	NUMBER OF PROPOSED PROJECTS			PROPOSED EXPENDITURES (in percent)			BUDGETED EXPENDITURES (in percent)		
RATIO	DIVISION REQUESTS (1963)	DIVISION REQUESTS (1964)	NEW BUSINESS (1963)	DIVISION REQUESTS (1963)	DIVISION REQUESTS (1964)	NEW BUSINESS (1963)	DIVISION REQUESTS (1963)	DIVISION REQUESTS (1964)	NEW BUSINESS (1963)
Ratio of Expenditure Initially Approved[a] by Laboratory Management to That Proposed by Department Manager									
0.00 to 0.39	3	5	6	13	38	42	0	16	0
0.40 to 0.79	7	4	4	24	24	32	19	20	15
0.80 to 1.19	20	8	4	47	36	23	61	53	72
1.20 to 1.59	2	0	1	2	0	3	3	0	13
1.60 and over	6	2	0	14	1	0	17	10	0
Total[b]	38	19	15	100	100	100	100	100	100
Ratio of Expenditure Finally Approved[c] by Laboratory Management to That Approved Initially by Them									
0.00 to 0.39	1	0	5	2	0	53	1	0	7
0.40 to 0.79	5	4	1	9	15	12	7	18	8
0.80 to 1.19	29	10	2	89	61	29	92	61	81
1.20 to 1.59	0	2	1	0	14	5	0	13	0
1.60 and over	0	2	1	0	10	1	0	7	4
Total[b]	35	18	10	100	100	100	100	100	100

[a] "Initially approved" means approved before the corporate management sets the laboratory's total budget.

[b] Because of rounding errors the percentages do not always sum to 100.

[c] "Finally approved" means approved after the corporate management has set the laboratory's total budget.

For example, among projects designed to satisfy division requests in 1963, about one-half experienced an alteration of 20 percent or more.

To explain the extent of the modification, we construct a model which is quite similar to that in section 2. We assume that the laboratory management, like the department manager, believes that equation (1) holds and that it too attempts to maximize expected profits subject to a budget constraint. However, it is recognized that the laboratory management has different information regarding X and a different value of λ. Thus,

$$C^*_M = ln\left[\frac{\beta X_M}{1 + \lambda_M}\right] / \beta, \tag{8}$$

where C^*_M is the level of expenditure approved by the laboratory manager, X_M is his estimate of X, and λ_M is the marginal expected discounted profits from an extra dollar spent in the laboratory. When equations (4) and (8) are combined, the percentage change made by the laboratory management in a given project's budget is

[13] These seventy-two applied R and D projects accounted for about $5 million of the laboratory's budget. The relevant data are in the first part of Table 3.1. However the rest of the data are also of interest. One might argue that we should be concerned with the level of expenditure finally approved by the laboratory, rather than C^*_M. However, we suspect that the same model would work in either case.

$$(9)\qquad \frac{C^*_M - C^*}{C^*} = \left\{ln\left[\frac{X_M}{X}\right] + ln\left[\frac{1+\lambda}{1+\lambda_M}\right]\right\}\left\{ln\left[\frac{\beta X}{1+\lambda}\right]\right\}^{-1}$$
$$= -\left\{ln\left[\frac{X_M}{X}\right] + ln\left[\frac{1+\lambda}{1+\lambda_M}\right]\right\}\{ln\,[1 - P^*]\}^{-1}$$

To test this model, we make two additional assumptions. First, the laboratory management is assumed to set the value of X_M on the basis of X_p, the project evaluation group's estimate of the discounted profits (net of division costs) if the project is successful.[14] In particular, suppose that

$$(10)\qquad X_M = B_0 X_p^{B_1} Z'',$$

where the B's are the same for all projects and Z'' is a random error term. Second, we assume that the deviations of the unobservable variable, $ln\,[(1 + \lambda)/(1 + \lambda_M)]$, from its average value, σ, can be treated as a random error term. Given these assumptions,

$$\frac{C^*_M - C^*}{C^*} = \left\{ln\left(\frac{\alpha_0}{B_0}\right) - \sigma + \alpha_1\,ln\,X_D - B_1\,ln\,X_p + Z'''\right\} / ln\,(1 - P^*)$$

where Z''' is a random error term. Finally, letting $ln\,(\alpha_0/B_0) - \sigma = \theta$, we have

$$(11)\qquad \left(\frac{C^*_M - C^*}{C^*}\right) ln\,(1 - P^*) = \theta + \alpha_1\,ln\,X_D - B_1\,ln\,X_p + Z'''.$$

To test the model, we collected data regarding X_p and C^*_M, as well as X_D, P^*, and C^*, for ten major projects,[15] and obtained least-squares estimates of θ, α_1, and B_1. The resulting regression,

$$(12)\qquad \left(\frac{C^*_M - C^*}{C^*}\right) ln\,(1 - P^*) = \underset{(0.88)}{-1.74} + \underset{(0.16)}{0.46}\,ln\,X_D - \underset{(0.17)}{0.22}\,ln\,X_p,$$

is quite consistent with the model, the regression coefficients having the expected sign and all but one being statistically significant. About 44 percent of the variation in the dependent variable can be explained. Also, in accord with the theory, the regression coefficient of $ln\,X_D$ in equation (7) does not

[14] The project evaluation group, like the division, made estimates of the profitability of a project. The department manager saw only the division estimates, whereas the laboratory management generally saw only the estimates prepared by the project evaluation group. Occasionally the laboratory management also saw the division estimates, which normally are less conservative than those of the project evaluation group. Of course, if the laboratory management's value of X_M is a linear function (in logs) of both X_p and X_D, it would not affect the results.

[15] Although the sample size is small, it should be kept in mind that these projects account for a large dollar expenditure. Note too that the costs involved in expanding the sample were quite large.

differ significantly from that in equation (12). Thus, the available data, although far less extensive than one would like, seem to support the model.[16]

4. Implications of the Results

To the extent that the results of this pilot study are borne out by further investigation, they have several implications. First, they suggest that, despite its obvious limitations,[17] the assumption of expected profit maximization is of use in understanding the allocation of R and D funds. Needless to say, this does not mean that various "noneconomic" factors, which are discussed in the following paragraph, are unimportant. However, it does mean that a substantial portion of the observed allocation of funds can be explained by a purely "economic" model of the most old-fashioned variety. This point is important in formulating public policy regarding R and D and in attempting to predict firm behavior. It makes a great deal of difference whether we assume that a firm's R and D portfolio is dictated largely by organizational factors and by variables other than profit which play an important role in the goal structure of the decision-maker—or whether we assume that profit expectations are at least as important as such "noneconomic" variables.[18]

Second, the results suggest that although profit expectations are important, they can explain only about half of the observed allocation of funds.[19] Interviews with company executives and scientists indicate that the following factors are particularly important in explaining the rest.[20] (1) Decisions made by company scientists reflect their scientific and professional goals, which

[16] Interviews with laboratory executives indicated that many of them agreed with the basic ideas behind the model. However, as would be expected, they did not necessarily agree with some of the simplifying assumptions. See Section 4.

[17] Some of the most obvious limitations of the model are its neglect of organizational variables, risk aversion, the nature of information flows, and the role of variables other than profit in the goal structure of the decision-maker.

[18] By "noneconomic" factors, we merely mean those not usually considered in the traditional theory of the firm. Note that the results suggest that the firm tends to invest in projects that it *believes* to be very profitable. Whether they are *actually* very profitable is unknown. See note 48. Also, these results pertain only to one firm. How typical this firm is in this respect is difficult to say.

[19] This assumes that P^* is held constant. Then, judging from the correlation coefficients in equations (7) and (12), about half of the variation in $\ln C^*$ or $\frac{C^*_M - C^*}{C^*}$ can be explained by these equations. Of course, it is not necessary to hold P^* constant. One could shift the terms in equations (7) and (12) that involve P^* to the other side of these equations, thus making C^* and C^*_M a function of P^* and the other variables. Then one could compute the percent of the variation in C^* or C^*_M that is explained. However, this is unlikely to change the results very much. Of course, the fact that C^* and P^* are determined at the same time does not make it wrong to use P^* as an independent variable which influences C^*.

[20] Of course, some of the unexplained variation is simply due to the fact that equation (1) is only a first approximation, that there is some interdependence among projects (see note 26), and that even if equation (1) held, the laboratory management's estimate of β would not always coincide with the department manager's. Moreover, in the long run, some of the factors discussed in this paragraph of the text may be quite consistent with profit maximization.

are not always consistent with the strictly commercial objectives of the firm.[21] (2) Intrafirm politics are important. For example, the laboratory management is influenced by the necessity of maintaining good relations with various operating executives who have an interest in promoting projects in their areas.[22] (3) At each level of the decision-making process, the estimates received by the decision-maker are increased, decreased, given less weight, or given more weight, depending on the source of the estimate and its probable accuracy.[23] This aspect of the decision-making process obviously increases the variance of Z and Z'', thus increasing the unexplained variation. (4) Because of the way projects tend to be judged within the firm, there is a tendency to concentrate on projects where it is likely that success can be claimed at the end of the year, even if other, more adventurous projects promise higher expected returns to the company. (5) Most important of all, some scientists and department managers are much more effective than others in arguing for their proposals, in mobilizing support for them in the operating divisions, and in "selling" them to the laboratory management.

Third, the results provide new insight into the intrafirm generation and use of information. Both the department manager and the laboratory management rely on others to generate the estimates of X which determine their decisions. Apparently, both tend to discount extremely high or extremely low estimates, α_1 and B_1 being significantly less than one.[24] Thus, both put themselves less far out on a limb with regard to any project. It would be very interesting to know whether the same kind of bias exists in other parts of the firm—and in other firms.

Finally, the limitations of the models in sections 2 and 3 should be noted. First, the assumption that π is the criterion function—and that equation (1) holds—is very rough.[25] Second, there is sometimes an interdependence among

[21] The conflicts between management and scientists, arising from different traditions and goals, have often been pointed out. (See, for example, S. Marcson, *The Scientist in American Industry* [New York: Harpers, 1960] and E. Mansfield, "Technical Change and the Management of Research and Development," *Technological Change and Economic Growth*, Ann Arbor: University of Michigan Press, 1965.) For example, the laboratory department manager is influenced by the talents, experience, and interests of the members of his department, the budgets for some projects being increased to accommodate the scientific interests of certain scientists or to give others something to do in a slack period.

[22] For one thing, these executives help to determine the laboratory budget.

[23] Cost estimates are generally taken more seriously than revenue estimates, and estimates by a distinguished scientist are taken more seriously than those by a novice.

[24] This assumes that, *on the average*, the estimate of X used by either of these decision-makers is approximately equal to that which is given to them. This seems reasonable.

[25] The limitations of equation (1) have been discussed in note 8. As for the use of π as the criterion function, it should be noted that this assumes that the decision-maker is neither a risk-averter nor a risk-lover. (For example, see H. Markowitz, *Portfolio Selection* [New York: Wiley, 1959]). Contrary to this assumption, it seems likely that many of the relevant decision-makers are risk-averters. Also, the use of π assumes that a project is either a "technical success" or a "technical failure" and that the latter results in a dead loss of C. This seems to be in accord with the firm's rating scheme, but it is clearly a naïve assumption. Also, the models in sections 2 and 3 do not apply to projects in which a small amount is invested in order to determine the feasibility of a certain approach. Of course, the firm does buy information in this way, but most of its bigger, more advanced projects are not viewed

projects which is ignored.[26] Third, as noted above, many important factors are omitted because of measurement problems—the interest, training, and experience of the research staff, "political" pressures and relations with the operating divisions, and the interests and ambitions of the department manager and the laboratory management. Fourth, the data underlying equations (7) and (12)—and the model—pertain almost entirely to the development end of the R and D spectrum.[27] Fifth, because of difficulties in obtaining satisfactory data, the empirical results in section 3 are based on a nonrandom sample of projects which may be biased somewhat in favor of the model.[28] Sixth, the models take as given the set of project proposals that the department manager or the laboratory management consider.[29] Seventh, the assumption that the department manager and the laboratory management have the same value of β is very rough.[30] Eighth, note once again that the models in

---

in this light. Instead, they are viewed as gambles with certain fairly definite commercial advantages as payoffs. Note that the models in sections 2 and 3 are static in the sense that they do not take account of the fact that projects may be postponed to future periods. In S. Marglin's terminology (*Approaches to Dynamic Investment Planning*, North-Holland, 1963), they are "myopic." However, in the present context, this is no deficiency. It seems to reflect the actual decision-making process.

[26] The probability that one project will be successful may depend on the probability that another is successful, because they draw on a common pool of manpower and/or the results of one project are helpful in carrying out the other. But if the relevant utility function were linear (as assumed), this would make no difference. However, one project's value of X may depend on that of another project, and this would make a difference.

[27] Obviously, the "noneconomic" variables are much more important at the research end of the R and D spectrum.

[28] We chose one-year projects to avoid ambiguities due to the use of a one-year budgetary period. In the case of longer-term projects, a cutback in next year's expenditures may represent only a transfer of funds from the next year to the more distant future, but no change in the project's total budget. Also, we were forced to include only those projects for which the necessary data could be obtained. It may well be that such projects are more likely than others to conform to the model.

[29] There is no explicit discussion of the factors influencing the generation of new project proposals and the decision to review the status of old ones. Obviously, the types of projects that are proposed (in a serious way) are influenced by the relative power of the technical and R and D personnel as contrasted with that of the operating divisions. See note 21. Most projects are reviewed when budgets and other such statements must be prepared. Of course, a particularly careful and quick review is triggered by a marked change in a project's progress, a large change in the resources it will require, or the occurrence of some event influencing strongly the importance of the project.

[30] Fortunately, if the laboratory management reduced all of the department managers' β's by the same proportion, the analysis would be unaffected. The department manager's estimate of β may be biased upward because of overoptimism or because he believes that it is good strategy in selling projects to underestimate their true costs. Suppose that the laboratory management divides these estimates of β for all projects by k. Then,

$$\frac{C^*_M - C^*}{C^*} = \frac{ln\left[\frac{\beta X_M}{k(1+\lambda_M)}\right] \Big/ k - ln\left[\frac{\beta X}{1+\lambda}\right]}{ln\left[\frac{\beta X}{1+\lambda}\right]},$$

and

$$-\left[\frac{C^*_M - C^*}{C^*}\right] ln\,(1 - P^*) = \frac{1}{k} ln\,X_M - ln\,X + U,$$

sections 2 and 3 are not meant to be detailed descriptions of the decision-making process. To repeat, the relevant managers do not go through the process of estimating the β's and maximizing π, but this is not important, so long as they act as if they do.

5. Characteristics of the Firm's Research and Development Portfolio

It is often claimed [51, 61] that the bulk of the R and D carried out by large corporations is relatively short-term, safe, and aimed at fairly modest advances in the state of the art. If true, this hypothesis has important implications regarding the role of the large firm in promoting technological change. How long-term were the projects carried out in the central research laboratory of this large firm? How great were the technical risks that were taken? To find out, we obtained data regarding the characteristics of seventy applied R and D projects carried out in 1963 or 1964. Executives of the firm seemed to regard these projects, which accounted for about $5 million of the firm's R and D budget, as being reasonably representative.[31] Tables 3.2, 3.3, and 3.4 show the results, based on equal weights for all projects, weights reflecting the size of the expenditures proposed for a project by the department manager, and weights reflecting the size of the expenditures budgeted for a project.

Table 3.2. *Number of Years Elapsing Between Beginning and Estimated Completion Date for Sixty-eight R and D Projects*

	NUMBER OF PROJECTS			PROPOSED EXPENDITURES (in percent)			BUDGETED EXPENDITURES (in percent)		
NUMBER OF YEARS	DIVISION REQUESTS (1963)	DIVISION REQUESTS (1964)	NEW BUSINESS (1963)	DIVISION REQUESTS (1963)	DIVISION REQUESTS (1964)	NEW BUSINESS (1963)	DIVISION REQUESTS (1963)	DIVISION REQUESTS (1964)	NEW BUSINESS (1963)
Less than 2.0	6	14	1	13	77	0	10	89	4
2.0 to 3.9	13	1	6	37	23	52	28	11	96
4.0 to 5.9	8	0	5	18	0	24	21	0	0
6.0 to 7.9	1	0	2	5	0	20	8	0	0
8.0 to 9.9	2	0	1	3	0	3	5	0	0
10.0 and over	8	0	0	23	0	0	28	0	0
Total[a]	38	15	15	100	100	100	100	100	100

NOTE: This is the completion date estimated by the department manager in 1963 or 1964 and not necessarily the completion date that was estimated when the project was begun.

[a] Because of rounding errors, the figures in the last six columns may not always sum to 100.

Most projects were expected to be completed in less than 4 years (Table 3.2), and the time interval between project completion and the application of the results was seldom expected to be more than 1 year (Table 3.3). Expectations

where U represents $ln\ [\beta/k(1 + \lambda_M)]/k - ln\ [\beta/(1 + \lambda)]$. Substituting the expressions in equations (5) and (10), we obtain an expression like equation (11), except that the regression coefficient of $ln\ X_p$ is B_1/k rather than B_1.

[31] However, they did seem to think that the sample of 1963 new business projects was somewhat biased toward short-range projects.

Table 3.3. *Estimated Number of Months Elapsing Between Completion of the Project and Application of the Invention, Twenty-seven R and D Projects*

	NUMBER OF PROJECTS		PROPOSED EXPENDITURES (in percent)		BUDGETED EXPENDITURES (in percent)	
NUMBER OF MONTHS	DIVISION REQUESTS (1963)	DIVISION REQUESTS (1964)	DIVISION REQUESTS (1963)	DIVISION REQUESTS (1964)	DIVISION REQUESTS (1963)	DIVISION REQUESTS (1964)
Less than 6.0	8	7	65	43	68	46
6.0 to 11.9	2	5	21	33	27	29
12.0 to 17.9	1	2	11	16	0	20
18.0 or more	1	1	3	8	3	5
Total[a]	12	15	100	100	100	100

NOTE: These estimates were made by the project evaluation group.

[a] Because of rounding errors, the figures in the last four columns may not always sum to 100.

of this sort are generally optimistic; according to company officials, the elimination of this bias would increase the figures in Tables 3.2 and 3.3 by about 50 percent. If one follows Hamberg [51] and defines short-term projects to be those taking 5 years or less, the bulk of the projects were short-term, even if this bias was eliminated.[32] However, this definition of a short-term

Table 3.4. *Estimated Probability of Technical Success for Seventy R and D Projects*

	NUMBER OF PROJECTS			PROPOSED EXPENDITURES (in percent)			BUDGETED EXPENDITURES (in percent)		
ESTIMATED PROBABILITY	DIVISION REQUESTS (1963)	DIVISION REQUESTS (1964)	NEW BUSINESS (1963)	DIVISION REQUESTS (1963)	DIVISION REQUESTS (1964)	NEW BUSINESS (1963)	DIVISION REQUESTS (1963)	DIVISION REQUESTS (1964)	NEW BUSINESS (1963)
0.90 to 1.00	15	6	7	33	24	54	36	24	21
0.80 to 0.89	12	8	1	39	63	6	45	58	7
0.70 to 0.79	6	3	5	17	12	34	10	18	72
0.60 to 0.69	0	0	0	0	0	0	0	0	0
0.50 to 0.59	4	0	1	8	0	4	5	0	0
Less than 0.50	1	0	1	3	0	2	4	0	0
Total[a]	38	17	15	100	100	100	100	100	100

NOTE: These estimates were made by the department managers before beginning the projects.

[a] Because of rounding errors, the figures in the last six columns may not always sum to 100.

project may be too stringent. If 3 years, rather than 5, were used as a cutoff point, a very substantial percentage of the projects would be long-term.

[32] These results do not seem atypical. We have similar data for several large petroleum and chemical firms. For three petroleum firms, projects that were expected to be completed and have an effect on profits in less than 5 years were 80, 65, and 50 percent of the total. For the chemical firm, such projects were 90 percent of the total. These figures pertain to 1964.

Most of the projects do not involve very great technical risks. In about three-fourths of the cases, the estimated probability of technical success exceeds 0.80 (Table 3.4). In part, this undoubtedly reflects the optimism so often found among professional researchers and the necessity to sell projects; but interviews with various executives of the firm indicate that if this bias were removed the bulk of the estimates would still be well above 0.50. Additional evidence to this effect is presented in the following section of this chapter, which describes the outcome of these projects. However, since technical success by no means insures commercial success, the total risks are considerably greater than is indicated by Table 3.4.[33]

To what extent does the laboratory management, in choosing among alternative proposals, discriminate against long-term, risky proposals? If the laboratory management tends to cut the proposed expenditures for long-term, risky projects more than for short-term, safe projects, the frequency distribution of projects, weighted by the size of the expenditures proposed by the department manager, should be more "long-term" and "risky" than the frequency distribution of projects, weighted by the size of the expenditures budgeted by the laboratory management. A comparison of the frequency distributions yields no evidence of such discrimination (Tables 3.2, 3.3, and 3.4). However, interview studies seem to suggest that the laboratory management does discriminate in this way.[34]

Finally, consider the estimated profitability of the R and D projects, if successful. Griliches [47] has shown that the rate of return from successful agricultural inventions was very high. Enos [34] has obtained similar results for petroleum refining. According to Table 3.5, the rate of return from these R and D projects would be about as high as those in agriculture and petroleum refining, if they too were successful. In practically all of the cases, the estimated rate of return exceeds 100 percent.[35]

[33] Of course, our results in no way contradict the finding of Chapter 1 that R and D is a risky process. Including the probability of commercial as well as technical failure, the risks are greater than in most other aspects of business. However, relative to far-reaching R and D projects, the technical risks taken in these projects do not seem to be very great.

These results are similar to those regarding several large petroleum firms. For three petroleum firms, projects with expected probabilities of success exceeding 0.75 were 60, 56, and 50 percent of the total. However, for the chemical firm, such projects were only 15 percent of the total.

[34] Obviously, the department managers may propose projects of this sort because they think that these are the types of projects that the laboratory management wants. The data in Tables 3.2, 3.3, and 3.4 cannot get at this sort of effect.

[35] Of course, this presumes that the project is successful. But, as Z. Griliches, in "Research Costs and Social Returns: Hybrid Corn and Related Innovations" (*Journal of Political Economy* [Oct. 1958]), points out, the rate of return on a successful project may be of interest, since "it may be useful to break down the probable rate of return into two components: the rate of return if the development turns out to be a success and the probability that it will be a success." Note too that the estimates in Table 3.5 seem higher than those for several large petroleum and chemical firms. In part, this may be due to the fact that the estimates in Table 3.5 are rates of return of the extra investment required to finish the projects. Naturally, the rate of return on the total investment is lower. Also, these figures are estimates of the private rate of return, not the social rate of return. (Griliches estimates the latter.) For crude estimates of the marginal rate of return in petroleum and chemicals, see Chap. 4.

To sum up, it appears that most projects are expected to be completed in 4 years or less, the results are expected to be applied only a few months later, and the estimated probability of technical success averages about 0.80. If successful, the rate of return from the investment in R and D is expected to

Table 3.5. *Estimated Rate of Return From R and D Projects If Technically Successful, Eleven Projects, 1963*

RATE OF RETURN (in percent)	DIVISION ESTIMATES	PROJECT EVALUATION GROUP ESTIMATES
	Number of Projects	
Less than 0	0	1
0 to 99	2	6
100 to 199	2	1
200 to 299	2	1
300 or more	5	2
Total	11	11

NOTE: The rates of return were computed on the basis of estimates generated by the divisions and the project evaluation group.

be very high. In interpreting these results, note that the central research laboratory of a large science-based firm tends to be more heavily committed to long-term, risky R and D than most industrial laboratories. Thus, the R and D portfolio described here is probably less conservative and short-term than that of most large firms.[36]

6. Outcome of the Research and Development Projects

Whereas previous sections were concerned with the allocation of the R and D budget and the characteristics of the projects that were attempted, the rest of this chapter is concerned with the outcome of these projects. What proportion were technically successful? What were the principal reasons for failure? To what extent did the actual expenditures on a project correspond with the budgeted expenditures? How accurate were the firm's predictions of the probability of success? Data are presented for the first time to help answer these questions.

To begin with, consider the proportion of the projects described in section 5 that turned out to be successful. By a "success," we mean a project in which, according to the judgment of the project evaluation group, the stated goals

[36] For further discussion of the role of the large firm in the creation, introduction, and acceptance of new technology, see E. Mansfield, *Economics of Technological Change*, and Chaps. 6–10 below.

were met in the expected length of time.[37] According to Table 3.6, about one-half of the projects were successful, the rest having been either delayed or dropped. Among those that were delayed, the average slippage factor (the

Table 3.6. *Outcome of Forty-five R and D Projects, 1963*

ITEM	TECHNICALLY SUCCESSFUL PROJECTS	REASONS FOR FAILURE[a]: MANPOWER DIVERTED TO HIGHER PRIORITY PROJECTS	OBJECTIVES CHANGED	UNFORESEEN TECHNICAL DIFFICULTIES	OTHER	TOTAL
Percent of:						
Projects	44	18	9	16	13	100
Budgeted expenditures	53	12	9	19	7	100
Actual expenditures	51	10	10	20	7	100
Slippage in schedule[b]:						
Mean	—	1.26	1.69	1.07	1.46	—
Standard deviation	—	0.36	0.62	0.19	—	—

NOTE: At the end of 1963, the project evaluation group rated all projects as "successful," "partially successful," or "unsuccessful." The last two classes are lumped together here.

[a] Note that these reasons were given by the project evaluation group, not the department manager. Some of the reasons included under "Other" are the inability to hire personnel, the fact that a key scientist quit, and the termination of a project because it no longer seemed justified commercially.

[b] The 1964 estimate of the time interval from the beginning to the end of the project, divided by the 1963 estimate.

revised time to completion divided by the original estimated time to completion) was about 1.30.

The proportion of failures is considerabiy higher than would be expected on the basis of the estimated probabilities of success in Table 3.4. Apparently, the primary reason is that about one-third of the projects were not carried out in the way the department manager proposed. In 18 percent of the proj-

[37] The project evaluation group reviewed the records of each of the projects carefully. In view of its position in the laboratory organization and the fact that the review was conducted for the laboratory management, there is no reason to suspect any appreciable bias in the group's findings. Of course, this is only one possible definition of success. Others could have been used. Moreover, we could have employed a richer classification than the simple dichotomy of success or failure. However, this procedure seems to be a sensible first step.

ects, manpower was diverted to higher priority projects;[38] in 9 percent, the objective of the project was changed; and in 13 percent, various miscellaneous factors (like a key scientist quitting the firm) prevented the project from being carried out as planned. In only 16 percent of the projects was failure due to unforeseen technical difficulties (Table 3.6). Note that there was no evidence that the other reasons for failure, e.g., diversion of manpower, existed after, or as a consequence of, unexpected technical difficulties.[39]

The changes in the staffing, nature, and objectives of the projects are reflected in the ratio of actual to budgeted expenditures (Table 3.7). In about

Table 3.7. *Ratio of Actual Expenditures to Final Budget Approved by Laboratory Management, Forty-five R and D Projects, 1963*

	NUMBER OF PROPOSED PROJECTS		BUDGETED EXPENDITURES (in percent)	
RATIO	DIVISION REQUESTS	NEW BUSINESS	DIVISION REQUESTS	NEW BUSINESS
0.00 to 0.39	2	0	1	0
0.40 to 0.79	19	2	45	17
0.80 to 1.19	14	2	50	77
1.20 to 1.59	2	1	2	7
1.60 and over	2	1	2	0
Total[a]	39	6	100	100

[a] Because of rounding errors, the figures in the last two columns may not always sum to 100.

one-half of the cases, actual expenditures were more than 20 percent below the budget, generally because manpower could not be obtained or had been diverted elsewhere. In about 15 percent of the cases, actual expenditures exceeded the budget by 20 percent or more, often because the objective of the project was changed.

These results are of interest for at least four reasons. First, they provide further evidence that the bulk of the applied R and D carried out in this laboratory involves relatively small technical risks. Including only those cases in which the project was carried out as planned, the chance of success averaged about 0.75. Second, they help to explain why the laboratory seldom uses parallel R and D efforts, which are based on the assumption that technical uncertainties are very great.[40] Third, they shed new light on the

[38] These projects with higher priorities often were government-financed projects on which the laboratory was working, or new projects suggested after the R and D portfolio was tentatively fixed.

[39] The project evaluation group was questioned very carefully concerning this point, which obviously is an important one in the interpretation of the data.

[40] More accurately, parallel R and D efforts rest on the assumption that a great deal can be learned relatively quickly regarding the costs of alternative development paths. However, if there is little uncertainty to begin with, there is likely to be relatively little learning either.

extent to which project failure and slippage are "man-made rather than caused by an ill-natured technology," [41] the importance of this question being pointed out by Klein [68], Scherer [140], and others.[42] Finally, they show for the first time the extent of the slippage in commercial R and D and the way it depends on the cause of the delay.[43]

7. Accuracy of Estimated Probabilities of Success

How accurate are the estimates of the probability of success made before the projects were started? This is a very important question because the techniques recommended by economists and operations researchers [16, 29, 38, 114, 130] for selecting R and D projects depend on the use of such estimates. If it is impossible to make reasonably accurate estimates, techniques which are relatively sensitive to errors in the estimates should be avoided, even if they have other features that seem desirable.[44]

To find out how accurate the estimates (that is, P^*) were, we construct a statistical discriminant function to predict whether or not a project was successful on the basis of its value of P^*. This function is optimal in the sense that, under specified conditions,[45] it minimizes the proportion of incorrect predictions. To estimate this discriminant function, we define a variable, U, where

$$(13) \qquad U = \begin{cases} -n_1/(n_1 + n_2) & \text{if the project is not a success} \\ n_2/(n_1 + n_2) & \text{if the project is a success,} \end{cases}$$

and n_1 is the number of successes and n_2 is the number of failures. Then we regress U on P^*, the result being

---

Note that these results do not contradict B. Klein's contention ("The Decision Making Problem in Development," *The Rate and Direction of Inventive Activity* [National Bureau of Economic Research, 1962]) that parallel R and D efforts should be carried out in military development. He is concerned with cases where large advances are being sought in the state of the art. Consequently, the uncertainties are undoubtedly much greater.

[41] F. Scherer "Comment," *The Rate and Direction of Inventive Activity* (National Bureau of Economic Research, 1962), p. 498.

[42] This question is important in interpreting the figures regarding failure rates and slippages in schedule, which are used repeatedly to measure the extent of the uncertainty in R and D. Judging from our results, these figures exaggerate considerably the extent of the technical risks involved.

[43] A. Marshall and W. Meckling (in "Predictability of Costs, Time and Success of Development," *The Rate and Direction of Inventive Activity* [National Bureau of Economic Research, 1962]) present data regarding slippages in military R and D, but no data have been available until now regarding commercial R and D. Apparently, the average slippage is somewhat smaller (1.3 vs. 1.5) than in military work. According to Table 3.6, the slippage tended to be greatest when the objective of the project has changed and least when technical difficulties are encountered.

[44] Unfortunately, the discussions of selection techniques seldom, if ever, contain any treatment of the sensitivity of the results to errors of this kind.

[45] This assumes that the a priori probability of a success is 0.5. See T. Anderson, *An Introduction to Multivariate Statistical Analysis* (New York: Wiley, 1958). These conditions are only approximately fulfilled, but if better approximations were used, the results would be about the same. See note 47.

(14) $$U = -0.6655 + \underset{(0.44)}{0.84}\,P^*.$$

Since projects that were not carried out as planned are obviously not relevant here, the results are based on the remaining twenty-five projects.[46]

If a project's value of U calculated from equation (14) exceeds zero, one should predict success; if not, one should predict failure. This rule makes optimal use of the data on P*, in the sense that it minimizes the probability of an incorrect prediction. How accurate is it? The fact that the regression coefficient of P* in equation (14) is statistically significant indicates that P* is of some use in predicting success or failure.[47] However, it does not seem to be of great use, since it predicts incorrectly in about one-third of the cases on which equation (14) is based.[47]

Of course, P* may be more useful as a predictor if various other factors are held constant. To find out how much improvement there is, we included the age and experience of the department manager and the type of project (cost reduction, product improvement, consulting) as additional explanatory variables. Unfortunately, there was no evidence that their inclusion made any difference, since none of these variables turned out to be statistically significant factors in the discriminant function.

To sum up, although there is a direct relationship between the estimated probability of success and the outcome of a project, it is not strong enough to permit very accurate predictions. Using a statistical discriminant function, the probability of error is about one-third.

[46] We include "partial successes" with "successes" here. It would make more sense to include them with "failures," but there are so few that it makes little difference which way they are treated. See note 47.

[47] Of course, good or bad performance is relative; what is good enough for one purpose is not good enough for another. One can obtain a probability of misclassification equal to $1 - P'$ (where P' is the a priori probability of technical success) merely by predicting that all projects will be successes, regardless of their estimated probability of technical success. The probability of misclassification based on this naive prediction (which makes no use of the estimated probabilities of technical success) can be employed as a standard against which to judge the probabilities of misclassification based on the discriminant function. The results depend on the a priori probability of technical success, but they do not vary greatly in the relevant range ($0.5 \leq P' \leq 0.75$). In this range, if "partial successes" are treated as failures, the probability of misclassification based on the discriminant function is only 10–30 percent lower than the probability of misclassification based on the naive prediction.

The probability of error cited in the text (about one-third) is an average of the probability of error based on (1) the classification of "partial successes" as successes and (2) the classification of "partial successes" as failures. If a similar average is used for alternative values of P' in the relevant range, the probability of misclassification based on the discriminant function is about 30 percent lower than the probability of misclassification based on the naive prediction. Thus, the probability of error generally is greater when "partial successes" are classified as failures rather than successes. Also, the results based on this average do not depend much on P' (at least in the relevant range).

8. Summary and Conclusions

The principal conclusions of this chapter are as follows: First, the assumption of expected profit maximization seems to be of use in explaining the allocation of funds among applied research and development projects. The size of the budget proposed for a project can be explained fairly well by a model which assumes that proposed spending is increased to the point where the increase in the probability of success is no longer worth its cost. The alterations made by the laboratory management in the proposed budget also can be explained fairly well by a similar model. However, these findings are extremely tentative, since the sample of projects is quite small and some other model may provide an even better explanation. Moreover, the model cannot explain about half of the variation in the allocation of funds.

Second, the following four factors seem to account for much of this unexplained variation. (1) Holding expected profit constant, safe projects are preferred over risky ones. (2) Some attempt is made to satisfy scientific as well as commercial objectives, the consequence being that some projects are justified more on the basis of scientific interest than expected profit. (3) Intrafirm politics are important; for example, projects differ considerably in the amount of pressure applied by operating executives to have them carried out. (4) Some scientists and department managers are much more effective than others in arguing for their proposals and in mobilizing support for them.

Third, a detailed description of the laboratory's applied research and development projects seems to indicate that most are expected to be completed in 4 years or less, and the results are expected to be applied only a few months later. The estimated probability of technical success averages about 0.80; and if the projects are successful, the estimated rate of return from the investment in the research and development is extremely high. These results are not very different from some preliminary findings pertaining to several large firms in the chemical and petroleum industries.

Fourth, failure rates and slippages in schedule, which are used repeatedly to measure the extent of the uncertainty in research and development, seem to exaggerate very greatly the extent of the technical uncertainties involved. About one-half of this laboratory's projects did not achieve their technical objectives on time. However, about two-thirds of these "failures" resulted from changes in objectives or the transfer of personnel to other projects. In only about one-third of these cases was there any evidence that "failure" was due, directly or indirectly, to technical difficulties. Thus, whereas the unadjusted failure rate would indicate that the average probability of technical success was about 0.50, it was really about 0.75.

Fifth, a comparison of the estimate of the probability of a project's technical success (made prior to the beginning of the project) with the outcome of the project indicates that such estimates have some predictive value, but not a great deal. A discriminant function based on these estimates predicted incorrectly in about one-third of the cases. This finding should be useful in the formulation and evaluation of various research and development budget allocation techniques, practically all of which rely on such estimates. Up to this point, no information was available regarding their accuracy.[48]

[48] Because of the difficulties in measuring the returns from an R and D project and the fact that too little time has elapsed to gauge even roughly the success of many of these projects, we do not and cannot present information on actual profit outcomes of projects or characteristics of projects that were successful in terms of return on investment. This is an important limitation, since technical success by no means implies commercial success.

CHAPTER 4

Returns from Industrial Research and Development and Determinants of the Rate of Technological Change

Previous chapters have referred repeatedly to the importance—and difficulty—of estimating the returns from industrial research and development. Having cursed the dark loudly and publicly,[1] my purpose here is to stimulate discussion of the problem and to light a few candles, limited though their power may be. The enormous difficulties, both conceptual and practical, in making estimates of this sort are all too obvious. Nonetheless, it is important that the task be begun, since business firms and government agencies badly need better measures of the returns from research and development.[2] This chapter presents a simple econometric model that may be of interest to researchers in this area. This model is still in the exploratory stages; it is by no means a practical tool.

In section 1, we show how the marginal rate of return from research and development expenditures can be estimated, assuming that a simple model of production holds and that all technological change is disembodied. In section 2, we present numerical results for a small number of manufacturing

[1] See E. Mansfield, "The Economics of Research and Development: A Survey of Issues, Findings, and Needed Future Research," *Patents and Progress* (Homewood, Ill.: Irwin, 1965).

[2] For agricultural research, Z. Griliches (in "Research Costs and Social Returns: Hybrid Corn and Related Innovations," *Journal of Political Economy* [Oct. 1958], and in "Research Expenditures, Education and the Aggregate Agricultural Production Function," *American Economic Review* [Dec. 1964]) has made some estimates of the rate of return. However, no estimates have been made for particular manufacturing industries or firms. For some estimates of pay-out periods, see McGraw-Hill, *Business Plans for Expenditures on Plant and Equipment* (annual).

firms and industries. Section 3 shows how such estimates can be obtained, when it is assumed that all technological change is capital-embodied, and section 4 presents numerical results based on this assumption. The findings are discussed in section 5, and section 6 investigates the determinants of the rate of technological change.

1. Disembodied Technological Change: Model

To begin with, we assume that all technological change is disembodied and that the production function for a particular firm is

$$(1) \qquad Q(t) = Ae^{a_1 t}\left[\int_{-\infty}^{t} e^{-\lambda(t-g)}R(g)\,dg\right]^{a_2} L^{\alpha}(t)K^{1-\alpha}(t),$$

where $Q(t)$ is the output rate (in 1960 prices) at time t, $L(t)$ is the labor input at time t, $K(t)$ is the stock of capital (1929 prices) employed at time t, $R(g)$ is the rate of expenditure (1960 prices) on R and D by this firm at time g, λ is the annual rate of "depreciation" of an investment in R and D, a_1 is the rate of technological change that would occur even if R and D expenditures (net of "depreciation") by this firm were to cease, a_2 is the elasticity of output with respect to cumulated past net R and D expenditures, A and α are parameters, and time is measured in years from 1960.

Five points should be noted regarding equation (1). First, it recognizes that the firm's efficiency depends on its previous R and D expenditures, as well as on the inventive activity of other firms and nonprofit institutions (the effect of the latter being incorporated in a_1).[3] Second, it allows for the possibility that an investment in R and D, like that in plant and equipment, "depreciates" over time, because of the obsolescence of research findings and designs.[4] Third, it makes the usual assumption that when time and the level of previous R and D expenditures—and consequently the level of technology—are held constant, there are constant returns to scale with respect to labor and capital. Fourth, A and α can vary from firm to firm, and a_1 and a_2 can vary from industry to industry.[5] Fifth, the little evidence that is available

[3]For other papers that assume that current output is a function of cumulated past R and D expenditures, see J. Minasian, "Technical Change and Production Functions" (unpublished paper presented at the fall 1961 meetings of the Econometric Society) and "The Economics of Research and Development," *The Rate and Direction of Inventive Activity* (Princeton, 1962). See also M. Kurz, "Research and Development, Technical Change, and the Competitive Mechanism," Institute of Mathematical Studies in the Social Sciences, Stanford, 1962. Apparently the model used in this chapter is the first to use a depreciation factor and to include "imported" technological change.

[4] One implication of this model, as it stands, is that the firm must do some R and D just to maintain its efficiency at a constant level. Since it takes some R and D simply to move along a given production function in response to changing factor prices, etc., this may not be too unrealistic. If one believes that this is not the case, he can set λ equal to zero. This makes little difference in the subsequent analysis.

[5] We assume that a_1 and a_2 are the same for firms within the same industry, but this assumption can be relaxed if one part of an industry seems to differ significantly from another part. In sections 2 and 4, we assume that a_2 is the same in chemicals as in petroleum.

seems to support equation (1). A study [109] of seventeen chemical firms indicates that this type of production function fits well and that the coefficients are generally statistically significant and of expected size. This evidence is hardly conclusive, but it is all we have.[6]

Next, assume that, for the period up to 1960,

$$R(g) = R_o e^{(\rho-\sigma)g}, \tag{2}$$

where R_o is the firm's R and D expenditures in 1960, ρ is the rate of increase of R and D expenditures in current dollars, and σ is the rate of price increase of R and D, as well as an allowance for noncomparability of R and D figures over time.[7] According to the available data, R and D expenditures in current dollars rose (approximately) exponentially in most firms during 1927–1960.[8] It seems reasonable to assume as a first approximation that R and D expenditures in constant dollars did so too.

Given these assumptions, what would have been the marginal rate of return from an additional amount spent in 1960 on R and D, if it is assumed that R and D expenditures after 1960 would continue to conform to equation (2)? If an additional expenditure of ϵ had been made on R and D in 1960, the output at a subsequent time t would have been

$$Q(t) = Ae^{a_1 t}\left(\frac{R_o e^{(\rho-\sigma)t}}{\rho-\sigma+\lambda} + \epsilon e^{-\lambda t}\right)^{a_2} L^{\alpha}(t)K^{1-\alpha}(t), \tag{3}$$

and

$$\left.\frac{\partial Q(t)}{\partial \epsilon}\right|_{\epsilon=0} = a_2 Ae^{(a_1-\lambda)t}\left(\frac{R_o e^{(\rho-\sigma)t}}{\rho-\sigma+\lambda}\right)^{a_2-1} L^{\alpha}(t)K^{1-\alpha}(t). \tag{4}$$

Thus, if $L(t) = L_o e^{lt}$ and $K(t) = K_o e^{kt}$, it follows that

---

Because these industries overlap and are similar in many respects, this may not be unreasonable. Of course, if more data were available, it would not be necessary to make this assumption.

[6] Note that Minasian's results (in "Technical Change and Production Functions") are based on the assumption that λ and σ equal zero and that his data on R and D expenditures do not go back very far. Moreover, because of the correlation between time and the sum of previous R and D expenditures, his estimates are not very precise when both of these variables are included in the production function. Nonetheless, his study, which is the only one of this sort that has been done to date, seems to point toward the use of equation (1) as a first approximation. For further comments see note 12.

[7] Without doubt, some of the reported increase in R and D expenditures has been due merely to more inclusive definitions of R and D and to exaggerated estimates made to impress stockholders and others.

[8] For the firms used below, the average coefficient of correlation during 1945–1958 between the natural logarithm of R and D expenditures during year t and t was 0.96. For the industries used below, the average correlation during 1927–1961 was 0.98. Judging by these very high coefficients, the assumption that R and D expenditures (in current dollars) increased exponentially is a good approximation.

$$(5)\quad \int_0^\infty \frac{\partial Q(t)}{\partial \epsilon}\bigg|_{\epsilon=0} e^{-rt}\,dt$$

$$= a_2 A \left(\frac{R_o}{\rho - \sigma + \lambda}\right)^{a_2 - 1} L_o^{\alpha} K_o^{1-\alpha}$$

$$\times \int_0^\infty e^{t[a_1 - \lambda - r + (a_2 - 1)(\rho - \sigma) - \alpha l - (1-\alpha)k]}\,dt$$

$$= a_2(\rho - \sigma + \lambda)\frac{Q}{R_o}$$

$$\cdot [r + \lambda - a_1 - (a_2 - 1)(\rho - \sigma) - \alpha l - (1 - \alpha)k]^{-1},$$

where Q is the 1960 value of $Q(t)$. We assume that the price of the product will remain constant from 1960 on, and that the supply of the R and D inputs to the firm is infinitely elastic. Consequently, since both $Q(t)$ and $R(g)$ are expressed in 1960 prices, we set the right-hand side of equation (5) equal to one, solve for r, and find that

$$(6)\quad r^1 = a_2(\rho - \sigma + \lambda)\frac{Q}{R_o} + a_1 - \lambda + (a_2 - 1)(\rho - \sigma) + \alpha l + (1 - \alpha)k,$$

where r^1 is the marginal rate of return from the extra expenditure in 1960 on R and D.

2. Disembodied Technological Change: Empirical Results

To illustrate how equation (6) can be used to estimate the marginal rate of return from R and D, we did a pilot study of ten chemical and petroleum firms and ten manufacturing industries. First, consider the case of the ten firms. From correspondence with the firms, annual reports, and *Moody's*, we obtained estimates for each firm of α, R_o, ρ, and Q. Assuming that the value of labor's marginal product is set equal to its wage, we used labor's average share of value-added in the firm (during 1946–1962) as an estimate of α. To estimate Q and R_o, we used data regarding the firm's value-added and R and D expenditures in 1960. To estimate ρ, we used the slope of the regression of the natural logarithm of the firm's R and D expenditures (current dollars) on time during 1945–1958.

The next step was to estimate $b' = a_1 + a_2(\rho - \sigma)$. Letting

$$(7)\quad \dot{V}/V = \dot{Q}/Q - \alpha\dot{L}/L - (1 - \alpha)\dot{K}/K,$$

it follows from equations (1) and (2) that

$$(8)\quad \ln V(t) = \text{constant} + b't.$$

For each firm, we obtained data regarding $Q(t)$, $L(t)$, and $K(t)$, for 1946–1962.[9] Letting $V = 1$ in 1946, we calculated $V(t)$ for each subsequent year, regressed $\ln V(t)$ on t, and used the resulting regression coefficient as an estimate of b'. Since the model applies only to periods when the firm is operating at full capacity, we omitted years when output was less than some previous year.[10]

Next, given an independent estimate of a_2, we estimated a_1, assuming that a_2 is the same for all of these firms but that a_1 may differ from industry to industry. Since

$$\hat{b}' = [a_1 - a_2\sigma] + a_2\rho + z, \tag{9}$$

where $\hat{b}'$ is our estimate of b' and z is the sampling error it contains, it follows that

$$a_1 = [\hat{b}' - a_2\rho] + a_2\sigma + z.$$

An independent estimate [109] of a_2, based on data for seventeen chemical firms, is .11, the standard error being .01. To check this estimate,[11] I regressed $\hat{b}'$ on ρ, and found that the regression coefficient, .12, was very close to .11, the difference not being statistically significant. Setting a_2 equal to .11,[12] the

[9] The data come from correspondence with the firms, annual reports, and *Moody's*. $Q(t)$ is value-added deflated by the wholesale price index in the industry. (It might have been preferable to use the Schultze-Tryon deflators, although, strictly speaking, they do not pertain to value-added.) $L(t)$ is the firm's total employment. To obtain $K(t)$, I multiplied the book value of the firm's assets in year t by the estimated ratio in year t of the deflated (1929 dollars) to the undeflated book value of the industry's assets. To obtain the latter ratio, I regressed Creamer's estimate of this ratio against time. See D. Creamer, "Postwar Trends in the Relation of Capital to Output in Manufactures," *American Economic Review* (May 1958). Note that the data on value-added contain an error, which was judged too small to be worth correcting. By mistake, they are net of R and D expenditures. However, since R and D expenditures are a very small and relatively constant percentage of value-added in these firms (about 5 percent), their omission is very unlikely to have an important effect on the findings. Note too that our estimate of α is very rough. We assume that labor's marginal product is set equal to its wage. (Of course, we could assume that physical capital's marginal product is set equal to its wage. The resulting estimate of α would be the same.) However, we cannot assume that the marginal products of both labor and capital are set equal to their wages, because this would leave nothing as compensation for the investment in R and D. Of course, if this assumption seems too unrealistic, there are other ways that α could be estimated.

[10] Of course, this has the unfortunate effect of reducing the number of observations, but there was no choice since no data were available concerning the rate of capacity utilization.

[11] We can obtain our own estimate of a_2 by regressing $\hat{b}'$ on ρ, since z should be uncorrelated with ρ. To allow for interindustry differences in a_1, one can allow the intercept to differ between industries. However, it turns out that these differences are not statistically significant.

[12] Note two things: First, since the two estimates are largely independent, it would be better to average Minasian's estimate of a_2 and ours rather than to use his alone. However, the two estimates are so close that this would make no difference. Second, it may be objected that Minasian's estimate of 0.11 is inappropriate here because it is based on the assumption that a_1 is zero. However, if we use the estimate he obtains when he relaxes this assumption, together with our results, we get almost exactly the same answer. Specifically, if we average his estimate (0.08) and ours (0.12), we get 0.10. Moreover, although each of these estimates has a substantial standard error, the average is statistically significant.

average value of $(\hat{b}' - a_2\rho)$ for our ten firms is .013, and there is no evidence that it differs significantly between industries.[13] Thus, omitting z,

$$a_1 = .013 + .11\sigma. \tag{10}$$

Finally, inserting these estimates of a_1 and a_2, together with the estimates of α, ρ, R_o, and Q, into equation (6) and assuming that $\alpha l + (1 - \alpha)k = .02$ for all firms, we obtain the estimates of r', given assumed values of σ and λ.[14] The results are shown in Table 4.1.

Table 4.1. *Estimates of b and b′ and Marginal Rates of Return from R and D Expenditures, Ten Chemical and Petroleum Firms*

	ESTIMATES		RATES OF RETURN: CAPITAL-EMBODIED				RATES OF RETURN: DISEMBODIED			
FIRM	b	b'	$\sigma = 0.04$ $\lambda = 0.04$	$\sigma = 0.04$ $\lambda = 0.07$	$\sigma = 0.07$ $\lambda = 0.04$	$\sigma = 0.07$ $\lambda = 0.07$	$\sigma = 0.04$ $\lambda = 0.04$	$\sigma = 0.04$ $\lambda = 0.07$	$\sigma = 0.07$ $\lambda = 0.04$	$\sigma = 0.07$ $\lambda = 0.07$
C1	0.0051	0.0035	0.12	0.18	0.03	0.12	0.04	0.04	0.04	0.04
C2	.0624	.0239	.25	.26	.25	.25	.02	.01	.02	.02
C3	.0200	.0260	.42	.46	.38	.42	.10	.11	.09	.10
C4	.0354	.0141	.33	.36	.29	.33	.14	.17	.12	.14
C5	.0534	[a]	.53	.58	.47	.53	.06	.06	.05	.06
P1	.0212	.0033	.24	.25	.20	.24	.13	.16	.11	.13
P2	.0594	.0191	.57	.72	.49	.57	.25	.33	.17	.25
P3	.0656	.0317	.64	.70	.56	.64	.51	.63	.40	.51
P4	.0947	.0107	.92	.99	.82	.92	.79	.96	.63	.79
P5	.0877	.0182	.73	.78	.67	.73	.31	.37	.26	.31

SOURCE: See sections 2 and 4.

[a] Less than zero.

Let's turn to the ten two-digit manufacturing industries in Table 4.2. The model in section 1 can be reinterpreted on an industry-wide basis, $Q(t)$, $L(t)$, etc., being regarded as industry aggregates rather than figures for individual firms. Based on this interpretation, estimates of α, ρ, Q, R_o, and b' were obtained for each industry by methods similar to those used for the firms.[15] However, since a_2 would be expected to vary considerably from one industry to another, we cannot estimate a_2 and a_1 in the way we did for the firms, unless data are available for a number of relatively homogeneous industries.

[13] The average value for petroleum firms was 0.015; for chemical firms, it was 0.011. The difference is not statistically significant.

[14] I assume that σ and λ equal 0.04 and 0.07. As we shall see, variations in σ and λ turn out to have relatively little effect on the results.

[15] $Q(t)$ is value-added (as given by the Census of Manufactures) deflated by the wholesale price index in the industry. To estimate value-added for 1946–1948 and 1962, we used the logarithmic regression during 1949–1961 of value-added on "National Income by Industry." The estimate of ρ is derived from the logarithmic regression of R and D expenditures on time during 1927, 1937, and 1953–1961. Estimates of R and D expenditures during 1927 and 1937 come from Y. Brozen, "Trends in Industrial Research and Development," *Journal of Business* (July 1960). The rest come from the National Science Foundation. The estimates of b′ come from B. Massell, "A Disaggregated View of Technical Change," *Journal of Political Economy* (Dec. 1961).

Having data for only a small number of industries, we rewrite equation (6) as

$$(11)\quad r' = \frac{Q}{R_o} b' \left(1 - x + \frac{a_2}{b'} \lambda\right) + b' - \lambda - \rho + \sigma + \alpha l + (1 - \alpha)k,$$

where $x = (a_1/b')$ is the proportion of the industry's technological change not due to its own R and D. Then I set a_2 equal to zero, insert the estimates of α, ρ, Q, R_o, and b' into equation (11), assume that $\alpha l + (1 - \alpha)k = .02$, and obtain the results shown in Table 4.2. It can easily be seen that these results are lower-bounds on the marginal rates of return.

Table 4.2. *Estimates of b and b′ and Lower-bounds on the Marginal Rates of Return from R and D Expenditures, Ten Manufacturing Industries*

			RATES OF RETURN							
	ESTIMATES		CAPITAL-EMBODIED				DISEMBODIED			
INDUSTRY	b	b′	$\sigma = .04$ $\lambda = .04$	$\sigma = .04$ $\lambda = .07$	$\sigma = .07$ $\lambda = .04$	$\sigma = .07$ $\lambda = .07$	$\sigma = .04$ $\lambda = .04$	$\sigma = .04$ $\lambda = .07$	$\sigma = .07$ $\lambda = .04$	$\sigma = .07$ $\lambda = .07$
						$x = .25$				
Chemicals	0.037	0.035	0.03	0.03	0.04	0.03	0.24	0.21	0.27	0.24
Machinery	[a]	.020	[b]	[b]	[b]	[b]	.01	−.02	.04	.01
Food	.047	.014	.58	.57	.58	.58	1.77	1.74	1.80	1.77
Paper	.034	.023	.26	.26	.27	.26	1.51	1.48	1.53	1.50
Instruments	.083	.010	.07	.07	.07	.07	−.07	−.10	−.04	−.07
Electrical equipment	.036	.037	.04	.04	.04	.04	.07	.04	.10	.07
Stone, clay, and glass	.015	.025	.08	.08	.08	.08	.81	.78	.84	.81
Furniture	.019	.010	.37	.37	.38	.37	2.49	2.46	2.52	2.49
Apparel	.030	.009	.98	.97	.99	.98	3.38	3.35	3.41	3.38
Motor vehicles	.086	.024	.04	.04	.05	.04	.01	−.02	.04	.01
						$x = .50$				
Chemicals	.037	.035	.02	.02	.02	.02	.14	.11	.17	.14
Machinery	[a]	.020	[b]	[b]	[b]	[b]	−.04	−.07	−.01	−.04
Food	.047	.014	.43	.42	.43	.43	1.15	1.12	1.18	1.15
Paper	.034	.023	.19	.18	.19	.19	.98	.95	1.01	.98
Instruments	.083	.010	.05	.05	.05	.05	−.08	−.11	−.05	−.08
Electrical equipment	.036	.037	.03	.02	.03	.03	.01	−.02	.04	.01
Stone, clay, and glass	.015	.025	.11	.05	.06	.11	.52	.49	.55	.52
Furniture	.019	.010	.27	.26	.27	.27	1.64	1.61	1.67	1.64
Apparel	.030	.009	.75	.74	.74	.75	2.23	2.20	2.26	2.23
Motor vehicles	.086	.024	.03	.03	.03	.03	−.04	−.07	−.01	−.04
						$x = .90$				
Chemicals	.037	.035	.00	.00	.00	.00	−.03	−.06	.00	−.03
Machinery	[a]	.020	[b]	[b]	[b]	[b]	−.10	−.13	−.07	−.10
Food	.047	.014	.11	.11	.11	.11	.17	.14	.20	.17
Paper	.034	.023	.04	.04	.04	.04	.13	.10	.16	.13
Instruments	.083	.010	.01	.01	.01	.01	−.10	−.13	−.07	−.10
Electrical equipment	.036	.037	.01	.01	.01	.01	−.08	−.11	−.05	−.08
Stone, clay, and glass	.015	.025	.01	.01	.01	.01	.05	.02	.08	.05
Furniture	.019	.010	.06	.07	.06	.06	.26	.23	.29	.26
Apparel	.030	.009	.22	.23	.22	.22	.38	.35	.41	.38
Motor vehicles	.086	.024	.01	.01	.01	.01	−.12	−.15	−.09	−.12

SOURCE: See sections 2 and 4.

[a] Due to sampling errors, this estimate violates the a priori constraint that $b' > 0$.

[b] No estimate can be made because the estimate of b' is negative.

3. Capital-embodied Technological Change: Model

Before discussing the results shown in section 2, let's consider an alternative model based on the supposition that technological change is capital-embodied,[16] not disembodied, and see the extent to which the results differ from those in section 2. For capital installed at time v which is still in existence at time t, the production function for a particular firm is assumed to be

$$(12) \qquad Q_v(t) = Ae^{a_1 v}\left[\int_{-\infty}^{v} e^{-\lambda(v-g)}R(g)\,dg\right]^{a_2} L_v^{\alpha}(t)K_v^{1-\alpha}(t),$$

where $Q_v(t)$ is the output rate (in 1960 prices) at time t from such capital and $L_v(t)$ is the rate of labor input being combined with this capital at time t. Of course, A, a_1, and a_2 are not the same as in sections 1 and 2.

Next, assume that all capital in a particular firm, regardless of vintage, depreciates at an annual rate of δ and that the firm's labor force is allocated efficiently among various vintages of capital. Given this assumption, one can show that

$$(13) \qquad Q(t) = AL^{\alpha}(t)\left\{e^{-\delta t}\int_{-\infty}^{t} e^{(a_1/(1-\alpha)+\delta)v} \cdot \left(\int_{-\infty}^{v} e^{-\lambda(v-g)}R(g)\,dg\right)^{a_2/(1-\alpha)} I(v)\,dv\right\}^{1-\alpha}$$

where $I(v)$ is the gross investment in plant and equipment (1929 prices) by the firm at time v. Of course, δ varies from industry to industry, but we assume that it is the same for all firms in the same industry. Finally, we assume once again that equation (2) holds up to 1960.

Given this alternative model, what would have been the marginal rate of return from an additional amount spent in 1960 on R and D, assuming that R and D expenditures after 1960 would continue to conform to equation (2)? If an additional expenditure of ϵ had been made on R and D in 1960, the output at a subsequent time t would have been

$$(14) \qquad Q(t) = AL^{\alpha}(t)\left\{e^{-\delta t}\left(\int_{-\infty}^{0} e^{\{\delta+[a_1+a_2(\rho-\sigma)]/[1-\alpha]\}v}I(v)\,dv\right) \cdot \left(\frac{R_o}{\lambda+\rho-\sigma}\right)^{a_2/(1-\alpha)} + e^{-\delta t}\int_{0}^{t} e^{[\delta+a_1/(1-\alpha)]v}\left[\frac{R_o e^{(\rho-\sigma)v}}{\lambda+\rho-\sigma} + \epsilon e^{-\lambda v}\right]^{a_2/(1-\alpha)} dv\right\}^{1-\alpha}.$$

Let $Q_0(t)$ be the output rate at time t if $\epsilon = 0$, and let

$$(15) \qquad J_o(t) = e^{-\delta t}\int_{-\infty}^{t} e^{[\delta+(a_1+a_2[\rho-\sigma])/(1-\alpha)]v}I(v)\,dv.$$

[16] Technological change is capital-embodied if, to be applied, all new technology has to be embodied in new plant and equipment. (See R. Solow, "Investment and Technical Progress," *Mathematical Methods in the Social Sciences* [Stanford, 1959].) No attempt will be made here to use a model which combines both disembodied and capital-embodied technological change.

That is, $J_0(t)$ is the "effective" stock of capital [128] [155] at time t if $\epsilon = 0$. Assuming that $I(v) = Ie^{iv}$ from 1960 on, it follows that

$$(16) \quad \left.\frac{\partial Q(t)}{\partial \epsilon}\right|_{\epsilon=0} = a_2(\lambda + \rho - \sigma)\frac{Q_o(t)}{J_o(t)}\frac{I}{R_o} \cdot \left[\frac{e^{[i-\lambda+a_1/(1-\alpha)+(a_2/(1-\alpha)-1)(\rho-\sigma)]t} - e^{-\delta t}}{\delta - \lambda + i + \frac{a_1}{1-\alpha} + \left(\frac{a_2}{1-\alpha} - 1\right)(\rho - \sigma)}\right]$$

where I is gross investment in plant and equipment in 1960. Assuming for simplicity that $Q_o(t)/J_o(t) = Q/J$,[17] where Q and J pertain to 1960, we have

$$(17) \quad \int_0^\infty \left.\frac{\partial Q(t)}{\partial \epsilon}\right|_{\epsilon=0} e^{-rt}\,dt = \frac{a_2(\lambda + \rho - \sigma)QI}{\left[\delta - \lambda + i + \frac{a_1}{1-\alpha} + \left(\frac{a_2}{1-\alpha} - 1\right)(\rho - \sigma)\right] JR_o} \times \left\{\int_0^\infty e^{[-r+i-\lambda+a_1/(1-\alpha)+(a_2/(1-\alpha)-1)(\rho-\sigma)]t}\,dt - \int_0^\infty e^{-(\delta+r)t}\,dt\right\}.$$

$$(18) \quad = a_2\frac{(\lambda + \rho - \sigma)}{(\delta + r)}\frac{Q}{J}\frac{I}{R_o} \cdot \left[r + \lambda - i - \frac{a_1}{1-\alpha} - \left(\frac{a_2}{1-\alpha} - 1\right)(\rho - \sigma)\right]^{-1}$$

where r is assumed to exceed $i - \lambda + \frac{a_1}{1-\alpha} + \left(\frac{a_2}{1-\alpha} - 1\right)(\rho - \sigma)$. Setting the right-hand side of equation (18) equal to one and solving for r, we have

$$(19) \quad (r^* + \delta)\left[r^* + \lambda - i - \frac{a_1}{1-\alpha} - \left(\frac{a_2}{1-\alpha} - 1\right)(\rho - \sigma)\right] = a_2(\lambda + \rho - \sigma)\frac{Q}{J}\frac{I}{R_o},$$

where r^* is the rate of return from the extra amount spent in 1960 on R and D.[18]

[17] Of course, this is only a simplifying assumption. It would be more realistic to assume that $Q_o(t)/J_o(t)$ will be an increasing function of t. If one is willing to make forecasts of the rate of change of this variable, it is easy to modify the analysis accordingly.

[18] In most cases, a good approximation of r^* is $\sqrt{a_2(\lambda + \rho - \sigma)\frac{IQ}{R_o J}}$. Note too that it may be possible eventually to obtain estimates of a_2 in various industries, thus enabling us to estimate the rate of return rather than merely to obtain a lower-bound. For example, if a_2 is the same in a number of industries, the technique in note 11 might be used on industry-wide data.

4. Capital-embodied Technological Change: Empirical Results

To illustrate how equation (19) can be used to estimate r^*, we again consider the ten chemical and petroleum firms and the ten manufacturing industries. To estimate I for each firm, we obtained data regarding its investment in plant and equipment (1929 dollars) in 1960. The slope of the regression of $\ln I(v)$ on v during 1946–1962 was used as a rough estimate of i. As an estimate of δ, we used the reciprocal of the length of life of plant and equipment in the industry, assuming that plant has a 45-year life and that the average life of equipment is given by the U.S. Treasury's 1962 *Depreciation Guidelines and Rules.* To estimate J, the deflated book value of the firm's 1960 fixed assets was multiplied by Phelps's estimate [128] of the ratio of "effective" to "old-style" capital in the business sector of the economy.[19]

The next step was to estimate $b = a_1 + a_2(\rho - \sigma)$ for each firm. If the assumptions in the previous section hold.

$$(20) \quad (1 - \alpha) \ln \{[W(t) - W(t-1) + \delta W(t)]/I(t)\} = \ln\left[A\left(\frac{R_o}{\lambda + \rho - \sigma}\right)^{a_2}\right] + bt,$$

where $W(t) = Q^{\frac{1}{1-\alpha}}(t)/L^{\frac{\alpha}{1-\alpha}}(t)$. Thus, in each firm, we calculated the value of the left-hand side of equation (20) for each year during 1946–1962, using the estimates of α and δ described above and the data provided by the firms regarding $Q(t)$, $L(t)$, and $I(t)$. Then we regressed this value on t and used the regression coefficient, $\hat{b}$, as an estimate of b. As in section 2, we omitted years when output was less than some previous year.

Next, we estimated a_1 and a_2, assuming that a_2 is the same for all firms but that a_1 may differ between industries. The regression of $\hat{b}$ on ρ was

$$(21) \quad \hat{b} = -0.024 + \underset{(0.242)}{0.673}\,\rho,$$

the correlation coefficient (adjusted for degrees of freedom) being 0.61, and there was no evidence that a_1 differs between the industries.[20] Thus, our estimate of a_2 was 0.673 and our estimate of a_1 was $-0.024 + 0.673\sigma$.

[19] Note four things: First, expenditures on plant and equipment are deflated by the average of the Commerce price indexes for producers durable equipment and construction. Second, it would be more realistic if δ depended upon the rate of technological change, rather than being exogenous. Third, we use Phelps's estimate of "effective" capital based on a 3 percent improvement rate and excluding inventories. (See E. Phelps, "The New View of Investment: A Neoclassical Analysis," *Quarterly Journal of Economics* [Nov. 1962].) The roughness of this procedure to estimate J need hardly be labored. Fourth, we use Creamer's data to deflate the book value of fixed assets. See Creamer, "Postwar Trends in the Relation of Capital to Output."

[20] No independent estimate of a_2 is available for the case where technological change is capital-embodied. Thus, we use the technique described in note 11.

Finally, inserting these estimates, together with the estimates of δ, α, ρ, i, I, R_o, Q, and J, into equation (19), we obtained estimates of r^*, given assumed values of σ and λ. The results are shown in Table 4.1.

Turning to the 2-digit manufacturing industries, we again reinterpret the model in terms of industry-wide aggregates. Estimates of δ, I, i, and J were obtained by methods similar to those used for the firms. Rewriting equation (19) as

$$(22) \quad (r^* + \delta)\left(r^* - \frac{b}{1-\alpha} + \lambda - i + \rho - \sigma\right) = b\left(1 - x + \frac{a_2\lambda}{b}\right)\frac{I}{R_o}\frac{Q}{J},$$

we set a_2 equal to zero, insert the estimates of δ, I, R_o, Q, J, i, α, and b into equation (22), and obtain the results shown in Table 4.2. Like the comparable results in section 2, they are *lower-bounds* on the rate of return, rather than estimates of the rate of return. Without data for more industries, we cannot obtain the latter.[21]

5. Discussion of the Estimates

In view of their roughness, it would be unwise to read too much into the results shown in Tables 4.1 and 4.2. However, several points are worth noting. First, like the results of previous studies, the estimates given in Table 4.1 tend to be very high. Among the petroleum firms, regardless of whether technological change was capital-embodied or disembodied, the marginal rates of return average from about 40 to 60 percent. Among the chemical firms, they average about 30 percent if technological change is capital-embodied, but only about 7 percent if it is disembodied. Even if the elasticity of supply to the firm of R and D inputs is less than infinite, as Machlup asserts, the rates of return remain high, so long as the elasticity remains within seemingly reasonable bounds.[22]

Second, when these results are compared with the few estimates of this

[21] The data regarding $Q(t)$ and R and D expenditures are described in note 15. $L(t)$ is total annual hours of labor as calculated from Census and Bureau of Labor Statistics figures. $I(t)$ is the deflated (1929 dollars) expenditures on plant and equipment, the estimate in current dollars coming from the Census of Manufactures and the deflator being the average of the Commerce price indexes for producer durables and construction. To estimate the Census investment figures for 1946–1948 and 1962, we used the regression of the Census investment data on the Office of Business Economics investment data. To see that the figures in Table 4.2 are lower-bounds, note that the right-hand side of equation (22) is directly related to a_2 and that the left-hand side of equation (22) is directly related to r^*. Thus, as we decrease a_2, r^* must decrease in order that the equation will hold.

[22] If true, these high marginal rates of return may persist because of the riskiness of R and D activities or because firms are ignorant of the true returns. For further discussion, see below. However, if there is a lag in the effect of R and D expenditures on the production function, our estimates in Tables 4.1 and 4.2 overestimate the true rate of return. If one knows the length of the lag, one can make the necessary adjustment quite easily. For example, in the case of disembodied technological change, it turns out that our estimates in Table 4.1 (with σ and λ equal to 0.04) are about 20 percent too high if the lag is 1 year. Obviously, this may be important. More work should be done to determine the extent of the lag and how it should be introduced into the analysis.

sort that have been published by other economists, it appears that although our estimates seem very high they are generally much lower than those obtained by others. However, when our results are compared with data we obtained from a number of large chemical and petroleum firms regarding the expected profitability of their current R and D projects, it appears that our estimates are considerably higher than those of the firms. To some extent, the latter difference is due to differences in concept and to sampling errors and other inadequacies in our estimates. But to the extent that this difference is real and our estimates are closer to the truth, it suggests an underinvestment in R and D, particularly among the petroleum firms.[23]

Third, the estimates given in Table 4.1 suggest that the marginal rate of return in chemicals was directly related to a firm's size, but in petroleum it was inversely related to it. To the extent that these differences, which seem to persist for all values of λ and σ in Table 4.1 and for capital-embodied and disembodied technological change, are real, they suggest that a transfer of R and D inputs from the smaller chemical firms to the largest ones and from the largest petroleum firms to the somewhat smaller ones might be desirable.[24]

Fourth, the lower-bounds given in Table 4.2 suggest that even if x is as large as 0.90 the marginal rate of return exceeded 15 percent in the apparel industry, regardless of whether technological change was disembodied or capital-embodied. If it was disembodied, the same was true for food and furniture. Although they are by no means unambiguous, these results suggest that there may have been an underinvestment in R and D in some of these industries.[25]

Fifth, whether technological change is capital-embodied or disembodied there is evidence of diminishing returns to scale from cumulated net R and D expenditures in chemicals and petroleum. If technological change is dis-

[23] For results of other studies, see Ewell, "Role of Research"; Griliches, "Research Expenditures, Education, and the Aggregate Agricultural Production Function" and "Research Costs and Social Returns"; Mansfield, *Economics of Technological Change*; and Chap. 3 of this volume. The data regarding the firms' expectations pertain to eight firms and were obtained by interviews and correspondence. These data, together with similar information which I have been collecting, will be used in a paper I am writing with Michael Hamburger on industrial R and D expenditures. To be as conservative as possible in computing the marginal rate of return in petroleum, I adjusted the data on R and D expenditures to include research related to "exploration and production," much of which is apparently omitted in the National Science Foundation definitions but included in an early American Petroleum Institute study. The adjustment was simply to inflate the reported expenditures (which follow the N.S.F. definition) in each year by the 1955 ratio of the total expenditures, according to the A.P.I. definitions, to the total expenditures, according to the N.S.F. definitions. Data are available only for 1955.

[24] Note three things: First, we assume in the text that, in a given industry, the R and D resources used by these firms are substitutable. Second, there seems to be some positive (but usually statistically nonsignificant) correlation between a firm's value of r' and its value of r^*. Third, because of the small number of firms in Table 4.1, the correlation between firm size and the marginal rate of return may not always be significant in a statistical sense.

[25] Of course, this presumes that x does in fact lie between 0.25 and 0.90 in all industries. This is reasonable, but by no means indisputable.

embodied, our results, like Minasian's [109], suggest that the average effect of a 1 percent increase in cumulated net R and D expenditures is a 0.1 percent increase in current output.[26] If technological change is capital-embodied, its effect is a 0.7 percent increase in current output.

6. Rates of Technological Change: Estimates and Determinants

It is important to note that b and b' can be interpreted as rates of technological change, as Solow has pointed out [154, 155]. Thus, Table 4.1 provides the first published estimates of the rates of technological change during the postwar period in particular firms, and equation (21) tells us that a firm's rate of capital-embodied technological change is directly related to the rate of growth of its accumulated R and D expenditures. If technological change is disembodied, the relationship is still direct, but not statistically significant. To what extent is a firm's rate of technological change also influenced by its ratio of R and D expenditures to sales or by its growth rate? To find out, I used each of these variables as an additional independent variable (besides ρ) in a regression with b (or b') as the dependent variable. The results provide no indication that either variable exerts an important influence on the rate of technological change. However, the sample is very small.[27]

Next, let us turn to the industry data; the estimates of the rate of capital-embodied technological change in various manufacturing industries are shown as b in Table 4.2. According to these estimates, technology advanced most rapidly in motor vehicles and instruments, and least rapidly in machinery, glass, and furniture. As would be expected, the estimated rate of capital-embodied technological change generally exceeds the estimated rate of disembodied technological change. However, there is relatively little correlation between them.[28] To what extent is an industry's rate of technological change, like a firm's, directly related to the rate of growth of its accumulated R and D expenditures? Regressing the estimates of b (or b') on ρ, we find that

$$\hat{b} = -0.044 + \underset{(0.273)}{0.668}\,\rho, \quad (\bar{r} = 0.62) \tag{23}$$

[26] See notes 11 and 12, and the corresponding part of the text.

[27] More specifically, we used as additional independent variables the ratio of R and D expenditures to sales in 1960 and the rate of growth of sales during 1945–1958. Neither was statistically significant. The rate of growth of profits during 1945–1958 was used too, with similar results.

These tests suggest that a firm's rate of growth, or its ratio of R and D expenditures to sales, is not responsible for the variation in both b (or b') and ρ. Nonetheless, it is possible that an identification problem exists in equation (21) and that b or b' influences ρ.

[28] For some alternative estimates of the rate of capital-embodied technological change in 2 digit manufacturing industries, see R. Solow, "Capital, Labor, and Income in Manufacturing," *The Behavior of Income Shares* (Princeton, 1964). For some discussion of the reasons why $\hat{b}$ will differ from $\hat{b}'$, and the sorts of differences that might be expected, see Phelps's "New View of Investment." Of course, another reason for differences is the existence of sampling errors in both $\hat{b}$ and $\hat{b}'$.

and

$$(24) \qquad \hat{b}' = -0.011 + \underset{(0.114)}{0.212}\, \rho. \quad (\bar{r} = 0.28)$$

Thus, contrary to Griliches's assertion [49], the estimated rate of technological change *is* positively correlated with the rate of increase of accumulated R and D expenditures, and this is true regardless of whether technological change is capital-embodied or disembodied.[29]

To what extent is an industry's rate of technological change influenced by its ratio of R and D expenditures to sales, its growth rate, or its concentration ratio? All of these variables have been suggested by economists [165] as important determinants of the rate of technological change. To answer this question, I assume that

$$(25) \qquad b' = \phi_0 + \phi_1\rho + \phi_2V + \phi_3C + \phi_4G + z',$$

where V is the industry's ratio of R and D expenditures to sales in 1951, C is the industry's concentration ratio in 1947, and G is the industry's rate of growth of value-added between 1948 and 1961. Based on data for all the industries for which Massell provides estimates of b', we find that the effects of V, C, and G are nonsignificant.[30]

In conclusion, the estimated rate of technological change, both at the level of the firm and the industry, is directly related to the rate of growth of its accumulated R and D expenditures. However, there is no evidence that such frequently used variables as the industry's (or firm's) ratio of R and D expenditures to sales, its rate of growth, or its concentration ratio exert an important influence on its rate of technological change.

7. Summary and Conclusions

The principal conclusions of this chapter are as follows: First, if the production function is Cobb-Douglas, if total past research and development expenditures, as well as labor and capital, are inputs, and if research and development expenditures have grown exponentially, one can obtain relatively simple expressions for the marginal rate of return from research and

[29] Despite the fact that a long time series is needed to obtain adequate estimates of ρ, Griliches uses only 1956–1962. Moreover, since this period is subsequent to most of the years to which the data on $Q(t)$ pertain, it is no wonder that he fails to obtain significant results. Note too that in the case of equation (24), the data pertain to all the industries for which Massell provides estimates of b' and for which data on ρ could be obtained. (However, for obvious reasons, the aircraft and missiles industry had to be excluded.) Thus, the results are not based only on the data in Table 4.2. Finally, I do not want to push the argument with Griliches too far. On the basis of the models in sections 1 and 3, there is no reason why there should be a positive correlation on the 2-digit level between b (or b') and ρ. Moreover, since the existing data are scant, the correlations in equations (23) and (24) may be due in part to chance. Nonetheless, if ρ is based on a long time series, the results do seem to be at variance with his remarks.

[30] The industries that are included are described in note 29.

development, whether technological change is capital-embodied or disembodied. If it is capital-embodied, the marginal rate of return is directly related to the elasticity of output with respect to total past research and development expenditures and the rate of investment, but inversely related to the amount spent in the past on research and development and the capital-output ratio. If it is disembodied, the marginal rate of return is directly related to the elasticity of output with respect to total past research and development expenditures and inversely related to the ratio of total past research and development expenditures to present output.[31]

Second, on the basis of these theoretical results, estimates of the marginal rates of return in 1960 were made for ten major chemical and petroleum firms, and lower-bounds for the marginal rates of return were estimated for ten manufacturing industries. Judging from the data for individual firms, the rate of return was very high in petroleum; in chemicals, it was high if technological change was capital-embodied but low if it was disembodied. In chemicals, the rate of return was directly related to a firm's size, but in petroleum it was inversely related to it. As for the industry data, the rate of return seems to have been relatively high (15 percent or more) in the food, apparel, and furniture industries.[32] Although these results are of interest, not much policy significance should be attached to them, for reasons discussed below.

Third, as a by-product of the analysis, estimates were obtained of the rate of technological change in these firms and industries. They suggest that a firm's rate of technological change is directly related to the rate of growth of its accumulated research and development expenditures, which would be expected on the basis of our model. The same kind of relationship holds on the industry level as well. The results also indicate that the postwar rate of capital-embodied technological change was relatively high in motor vehicles and instruments and relatively low in machinery, stone, glass, and clay. However, there was little correlation between these estimates and those based on the assumption that technological change was disembodied.

In conclusion, these results should be viewed with considerable caution,

[31] Whether technological change is disembodied or capital-embodied, the rate of return depends too on the rate of "depreciation" of an investment in R and D and the rate of price increase for R and D. If it is disembodied, it depends on the rate of increase of expenditures on plant and equipment and the rate of "depreciation" of plant and equipment. None of these variables has a very great effect on the results. They do not vary over a wide range, and the results are not very sensitive to their variation. The most important variables seem to be those described above.

[32] Of course, this assumes that $x \leq 0.90$. Moreover, the results depend substantially on whether or not technological change is assumed to be capital-embodied or disembodied. Also, the rate of return in other industries may be higher than in these industries. Table 4.2 contains only lower-bounds. Returning to one of the themes of Chapter 1, these estimates of the rate of technological change—and the others considered in this chapter—are very rough. Moreover, there are important problems in distinguishing between capital-embodied and disembodied technological change. See D. Jorgensen, "The Embodiment Hypothesis," *Journal of Political Economy* (Feb. 1966) and E. Mansfield, *Economics of Technological Change*, Chap. 2.

for at least four reasons. First, they are based on a number of highly simplified assumptions—that uncertainty can be ignored, that technological change is cost-reducing, that all technological change is neutral, that the production function is Cobb-Douglas, and that capital is "putty," not "hard-baked clay." [33] [129] Second, the estimates in Tables 4.1 and 4.2 contain substantial sampling errors. Third, they are incomplete estimates of the social rate of return, since they do not take account of the effects of increased research and development expenditures in one industry or firm on productivity in another industry or firm. The social rates of return may be higher than ours. Fourth, although it is easy to include lags in the effect of research and development expenditures on the production function, as well as a finite elasticity of supply of research and development inputs to the firm, this was not done because of the lack of relevant data.[34]

[33] Note three things: First, the assumption that all technological change is cost-reducing is, of course, only a crude approximation, as has often been pointed out. This helps to explain our choice of the petroleum and chemical industries, both of which direct a great deal of their R and D at new processes. Second, in view of the importance of uncertainty in R and D, it may seem extremely unrealistic to assume a deterministic relationship between cumulated R and D expenditures and output. However, appealing to the law of large numbers, the uncertainties for an entire firm or industry may be much smaller than for a particular project. Third, we have not mentioned the effects of education and other improvements in labor on output levels. In the case of disembodied technological change, these effects are included in a_1.

[34] See note 22.

Part III

Technological Innovation

CHAPTER 5

Size of Firm, Market Structure, and Innovation

An invention, when applied for the first time, is called an innovation. Traditionally economists have stressed the distinction between an invention and an innovation, on the ground that an invention has little or no economic significance until it is applied. This distinction becomes somewhat blurred in cases like Armco's continuous wide-strip mill, where the inventor and the innovator are the same firm. Under these circumstances, the final stages of development may entail at least a partial commitment to a market test. However, in many cases the inventor is not in a position to—and does not want to—apply his invention, because his business is invention, not production, or because he is a supplier, not a user, of the equipment embodying the invention, or for some other reason. In these cases, the distinction remains relatively clear-cut.[1]

Regardless of whether the break between invention and innovation is clean, innovation is a key stage in the process leading to the full evaluation and utilization of an invention. The innovator—the firm that is first to apply the invention—must be willing to take the risks involved in introducing a new and untried process, good, or service. By obtaining needed information regarding the actual performance of the invention, the innovator plays a vital social role.[2] In this chapter, we take up the role of the large firm in innovation. As in parts of Chapters 2, 3, and 4, we are concerned here with the classic question:[3] What are the relative merits of various market structures—and the

[1] In cases where the invention is a new piece of equipment, both the firm that is first to sell the equipment and the firm that is first to use it may be regarded as innovators. The first user is important because he, as well as the supplier, often takes considerable risk.

[2] For further discussion of the innovation process, see E. Mansfield, *The Economics of Technological Change* (New York: Norton, 1968), Chap. 4.

[3] For earlier work, see J. Schumpeter, *Business Cycles* (New York: McGraw-Hill, 1939); J. Galbraith, *American Capitalism* (Boston: Houghton Mifflin, 1952); A. Kaplan, *Big Enterprise in a Competitive System* (Washington, D.C.: The Brookings Institution, 1954); D. Lilienthal, *Big Business: A New Era* (New York: Harpers, 1953); W. MacLaurin, "Technological Progress in Some American Industries," *Quarterly Journal of Economics* (Feb. 1953); H. Villard, "Competition, Oligopoly, and Research," *Journal of Political Economy* (Dec. 1958); E. Mason, "Schumpeter on Monopoly and the Large Firm," *Review of Economics and Statistics* (May 1951); W. Mueller, "A Case Study of Product Discovery and Innovation Costs," *Southern Economic Journal* (July 1957); G. Nutter, "Monopoly, Bigness, and Progress, "*Journal of Political Economy* (Dec. 1956); G. Stigler, "Industrial

role of corporate giants—in promoting technological change and the rapid utilization of new technology?

First, we investigate the extent to which the largest firms in several important industries have been the innovators. Second, we outline a simple model that seems to be useful in explaining why these giants accounted for a disproportionately large share of the innovations in some industries but not in others. Third, we try to estimate whether fewer innovations would have been introduced if the largest firms had been broken up. Fourth, we try to determine whether the smaller firms do less innovating, relative to the larger firms, than in the past. Fifth, we show how, under certain circumstances, one can make crude estimates of the effect of a change in market structure on how rapidly inventions made outside the industry where they are applicable will be applied. Of course, the results are by no means free of difficulties, since the basic models are often convenient first approximations, and the data are often rough. Nonetheless, they should be useful to those interested in the role of the very large firm in the application of new ideas.

1. Innovation and Size of Firm

Several decades ago, Schumpeter [149] challenged the then-prevailing view and asserted that in recent times innovations have been carried out primarily by very large firms. More recently, Galbraith [40], Kaplan [62], Lilienthal [73], and Villard [167] have taken much the same position, resting their case in considerable part on the following three arguments: First, the costs of innovating are so great that only large firms can now become involved. Second, projects must now be carried out on a large enough scale so that successes and failures can in some sense balance out. Third, for innovation to be worthwhile, a firm must have sufficient control over the market to reap the rewards.[4]

Organization and Economic Progress," *The State of the Social Sciences* (Chicago: University of Chicago Press, 1956); J. Bain, *Pricing, Distribution, and Employment* (New York: Henry Holt, 1953); Y. Brozen, "Invention, Innovation, and Imitation," *American Economic Review* (May 1951); J. Robinson, *The Rate of Interest* (New York: Macmillan, 1952); G. Stocking, *Testimony Before Subcommittee on Study of Monopoly Power*, Judiciary Committee, House of Representatives, 1950; J. Jewkes, D. Sawers, and R. Stillerman, *The Sources of Invention* (New York: St. Martin's Press, 1958); I. Stelzer, "Technical Progress and Market Structure," *Southern Economic Journal* (July 1956); J. Schmookler, "Bigness, Fewness, and Research," *Journal of Political Economy* (Dec. 1959); T. Scitovsky, "Economic Theory and the Measurement of Concentration," *Business Concentration and Price Policy* (Princeton, 1955); W. Fellner, "The Influence of Market Structure on Technological Progress," *A.E.A. Readings in Industrial Organization and Public Policy* (Homewood, Ill.: Irwin, 1958); and J. Schumpeter, *Capitalism, Socialism, and Democracy* (New York: Harper and Row, 1942).

[4] See E. Mansfield, *Monopoly Power* and *Economic Performance* (New York: Norton, 1964). Some of these writers were concerned primarily with invention, not innovation. But if the largest firms are likely to carry out a disproportionately large share of the inventions, they are even more likely to carry out a disproportionately large share of the innovations, since innovation is generally more risky and costly than invention.

This position has been questioned by Mason [107] and others, on the ground that there is no evidence that a disproportionately large share of the significant innovations has been carried out by very large firms. However, neither Mason and his followers nor Schumpeter, Galbraith, *et al.*, have carried out the empirical studies that are needed to settle the question. Thus, the argument continues.[5]

This section presents empirical findings regarding three basic industries—iron and steel, petroleum refining, and bituminous coal. Because of the difficulties involved in obtaining fairly complete data, it was impossible to include a larger number of industries. The findings are very rough, both because these industries may not be entirely representative and because of the obvious difficulties any study of this sort must face. Nonetheless, they should help to fill an important gap.

To obtain the data, trade associations and trade journals in each industry were asked to list the important processes and products first introduced in the industry since 1918. They were also asked to rank them by importance. Having obtained these lists, we consulted technical journals and corresponded with various firms inside and outside the industry to determine which firm first introduced each innovation commercially and when this took place. This information could be gotten for about 80 percent of the innovations. The results are contained in Tables 5.1, 5.2, and 5.3.[6]

Next, we obtained data regarding the size of each firm in each of the industries—the ingot capacity of each iron and steel firm in 1926 and 1945, the daily crude capacity of each petroleum firm in 1927 and 1947, and the

[5] Empirical studies of this sort have been recommended by E. Mason ("Schumpeter on Monopoly and the Large Firm," *Review of Economics and Statistics* [May 1951]) P. Hennipman ("Monopoly: Impediment or Stimulus to Economic Progress," *Monopoly and Competition and Their Regulation* [New York: Macmillan, 1954]), and T. Scitovsky ("Economic Theory and the Measurement of Concentration," *Business Concentration and Price Policy* [Princeton, 1955]); but very little has been done in this area. Of course, there are considerable—and obvious—difficulties in defining a particular innovation, in singling out the innovators, and in gauging the relative importance of various innovations. In view of these difficulties, any empirical study must be arbitrary in some respects and the results can only be rough approximations.

[6] The distinction between a process and a product innovation may sometimes be blurred because a new technique that reduces the cost of some product may also alter it somewhat. In such cases, we asked the respondent to make a judgment as to whether the alteration was great enough for it to be considered a new product. For further comments on the source of Tables 5.1, 5.2, and 5.3, see Appendix B and Chap. 6. Note that the first installation of a new process, or the first introduction of a new product, may not be the most important in many technical and financial respects. Also, a number of firms sometimes are among the first, and it is difficult to decide which one had priority. In such cases, the credit is split equally among the firms. In a few cases where there were more than one innovator, firms that were not primarily steel firms or ingot producers were among the innovators, but for the reasons discussed in Appendix B, they are not listed in Tables 5.1, 5.2, and 5.3. For example, Babcock and Wilcox is excluded in the case of continuous casting. Also, there is occasionally a question about whether a particular installation was commercial or experimental. For example, it can be argued that the installations of continuous casting by Allegheny Ludlum, Babcock and Wilcox, and Republic were experimental. If so, it would not change our results in any important way.

Table 5.1. *Innovations and Innovators, Iron and Steel Industry, 1919–1938 and 1939–1958*

INNOVATION	INNOVATOR
1919–1938	
Austempering	U.S. Steel
Continuous wide-strip mill	Armco
Continuous pickling	Wheeling
Continuous galvanizing	Armco
Mechanical scarfing	U.S. Steel
Multiple block wire drawing	U.S. Steel
Automatic operation of open hearth	Laclede
Electronic inspection of tin plate	Jones and Laughlin
Electrically welded pipe	Republic
Dolomite gun	Donner
Coreless induction electric furnace	Heppenstall
Electrolytic tin plate	U.S. Steel
High-strength alloy steels	U.S. Steel
Low-tungsten high-speed tool steel	Universal and Cyclops
Grain-oriented electric steel	Allegheny
Nonaging steel	Armco
18-8 stainless steel	Allegheny
Nitriding steels	Ludlum
Boron-treated steels	U.S. Steel
Valve steels	Ludlum
5 percent chrome hot-work tool steels	Braeburn
Continuous annealing[a]	Crown Cork and Seal
Continuous butt-weld pipe[a]	Fretz-Moon
High-temperature alloys[a]	Timken Roller Bearing
Nickel-bearing electrical steel[a]	Western Electric
1939–1958	
Stretch process for hot reducing tubes	U.S. Steel
All-basic open-hearth furnace	U.S. Steel
Ultrasonic testing	Republic
High top pressure blast furnace	Republic
Jet tapper	Republic
Differential coating of tin plate	National
Electric eye for Bessemer turndown	Jones and Laughlin
Vacuum melting	Allegheny Ludlum; Crucible
Continuous casting	Allegheny Ludlum; Crucible
Vacuum degassing (pouring)	Bethlehem; U.S. Steel
L-D oxygen process	McLouth
Oxygen lancing of open hearth	Bethlehem; National; Jones and Laughlin; Republic; U.S. Steel

Table 5.1. *Continued*

INNOVATION	INNOVATOR
Automatic programing of mills	Allegheny Ludlum
Killed Bessemer steel	U.S. Steel
Precipitation-hardening stainless steel	Armco
Manganese stainless steel	Allegheny; Republic
Aluminum-clad sheets	Armco
Titanium-treated enameling steels	Inland
Columbium-treated high-strength steel	National
Extra low-carbon stainless steel	U.S. Steel; Armco; Allegheny Ludlum; Crucible; Republic
Closed television circuits[a]	Babcock and Wilcox
Carbon lining for blast furnace[a]	Interlake Iron
Hot extrusion[a]	Babcock and Wilcox
Sendzimir cold mill[a]	Signode Steel Strap

SOURCE: See Appendix B.

[a] Innovations excluded from Tables 5.4 to 5.8 because innovator had no ingot capacity or because it was engaged primarily in another business.

production of each coal firm in 1933 and 1953. These data were collected primarily from government documents and trade directories, but in a few cases they had to be obtained directly from the firms. Next, we determine how many of these innovations were first introduced by the largest four firms in each industry. Since the recent situation probably differed from that in the prewar era, innovations that occurred during 1939–1958 were separated from those that occurred during 1919–1938.

Do the results indicate that the largest firms introduced a disproportionately large share of the innovations? Of course, this depends on what one means by a disproportionately large share. But if the largest firms devoted the same proportion of their resources as smaller firms both to inventive activity and to the testing and development of other people's ideas, if they could obtain applicable results as easily, and if they were as efficient and as quick to apply the results, one would expect their share of the innovations to equal their share of the market.[7] According to the rather crude measurements in Table 5.4, the largest four coal and petroleum firms carried out a larger number of innovations than this, but the largest four steel producers carried out fewer. Thus, if the Schumpeterian hypothesis is taken to mean that the largest firms accounted for a larger share of the innovations than

[7] It could also be that they devote more of their resources to inventive activity and less to testing and trying out inventions made by outsiders, and that fewer innovations are produced per dollar of expenditure on the former activity. This is difficult to check. But there is no evidence that it was the case in the steel industry—the only case in which their share of the innovations was less than their share of the market.

Table 5.2. *Innovations and Innovators, Petroleum-refining Industry, 1919–1938 and 1939–1958*

INNOVATION	INNOVATOR
1919–1938	
Burton-Clark cracking	Standard (N.J.)
Dubbs cracking	Shell
Fixed-bed catalytic cracking	Sun
Propane deasphalting of lubes	Union
Solvent dewaxing of lubes	Indian
Solvent extraction of lubes	Associated
Catalytic polymerization	Shell
Thermal polymerization	Phillips
Alkylation (H_2SO_4)	Standard (N.J.)
Desalting of crude	Ashland
Hydrogenation	Standard (N.J.)
Pipe stills and multidraw towers	Atlantic
Delayed coking	Standard (Ind.)
Clay treatment of gasoline	Barnsdall
Ammonia	Shell
Ethylene	Standard (Ind.)
Propylene	Standard (N.J.)
Butylene	Standard (N.J.)
Methanol	Cities Service
Isopropanol	Standard (N.J.)
Butanol	Standard (N.J.)
Aldehydes	Cities Service
Naphthenic acids	Standard (Calif.)
Cresylic acids	Standard (Calif.)
Ketones	Shell
Detergents	Atlantic
Odorants	Standard (Calif.)
Ethyl chloride	Standard (N.J.)
Tetraethyl lead as antiknock agent[a]	Refiners
Octane numbers scale[a]	Ethyl
1939–1958	
Moving-bed catalytic cracking	Socony
Fluid-bed catalytic cracking	Standard (N.J.)
Catalytic reforming	Standard (Ind.)
Platforming	Old Dutch
Hydrogen-treating	Standard (N.J.)
Unifining	Union; Sohio
Solvent extraction of aromatics	Standard (N.J.)
Udex process	Eastern State
Propane decarbonizing	Cities Service

Table 5.2. *Continued*

INNOVATION	INNOVATOR
Alkylation (H Fl)	Phillips
Butane isomerization	Shell
Pentane and hexane isomerization	Standard (Ind.)
Molecular sieve separation	Texaco
Fluid coking	Standard (N.J.)
Sulfur	Standard (Ind.)
Cyclohexane	Phillips
Heptene	Standard (N.J.)
Tetramer	Atlantic
Trimer	Atlantic
Aromatics	Standard (N.J.)
Paraxylene	Standard (Calif.)
Ethanol	Standard (N.J.)
Butadiene	Standard (N.J.); Shell
Styrene	Shell
Cumene	Standard (Calif.)
Oxo alcohols	Standard (N.J.)
Dibasic acids	Standard (Calif.)
Carbon black (oil furnace)	Phillips
Glycerine	Shell
Synthetic rubber	Standard (N.J.)
Ethylene dichloride	Standard (N.J.)
Diallyl phthalate polymers	Shell
Epoxy resins	Shell
Polystyrene	Cosden
Resinous high-styrene copolymers	Shell
Polyethylene	Phillips

SOURCE: See Appendix B.

[a] Innovations excluded from Tables 5.4 to 5.8 because innovator had no crude capacity or because it was engaged primarily in another business.

of the market, it seemed generally to hold in petroleum and coal, but not in steel.[8]

[8] The unweighted data in Table 5.4 suffer from the lack of a clear-cut way to define an innovation and gauge its importance. Conceivably, some of these innovations could be regarded as a set of separate innovations—not one. If they were, the results using unweighted data would depend on how many elements were recognized in each case. The weighted data should eliminate this problem, but the weights are obviously very crude. In addition, the lesser—and some important—innovations are excluded altogether; and hence sampling errors and perhaps biases (despite the opinions quoted in Appendix B) are present. Finally, note that in Table 5.4 there is much less variation in the share of innovations accounted for by the four largest firms than in their share of capacity or output.

Table 5.3. *Innovations and Innovators, Bituminous Coal Preparation, 1919–1938 and 1939–1958*

INNOVATION	INNOVATOR
1919–1938	
Simon-Carves washer	Jones and Laughlin; Central Indiana
Stump air-flow cleaner	Barnes
Chance cleaner	Rock Hill
"Roto Louvre" dryer	Hanna
Vissac (McNally) dryer	Northwestern Improvement
Ruggles-Cole kiln dryer	Cottonwood
Rheolaveur	American Smelting
Menzies cone separator	Franklin County
Deister table	U.S. Steel
Carpenter dryer	Colorado Fuel and Iron
Froth flotation	Pittsburgh
1939–1958	
Raymond flash dryer	Enos
CMI drying unit	Hanna
Link-Belt separator	Pittsburgh
Bird centrifugal filter	Consolidation
Baughman "Verti-Vane" dryer	Central Indiana
Vissac Pulso updraft dryer	Northwestern Improvement
Link-Belt multi-louvre dryer	Diamond; Elkhorn; Bethlehem; Eastern Gas and Fuel
Eimco filter	United Electric
Dorrco fluosolids machine	Lynnville
Parry entrainment dryer	Freeman
Heyl and Patterson fluid bed dryer	Jewell Ridge
Feldspar type jig	Northwestern Improvement
Bird-Humboldt centrifugal dryer	Clinchfield
Wemco Fagergren flotation unit	Hanna; Sevatora; Diamond
Continuous horizontal filter	Island Creek
Cyclones as thickeners[a]	Dutch State Mines

SOURCE: See Appendix B.

[a] Omitted from Tables 5.4 to 5.8 because innovator was not a domestic firm.

2. A Simple Model

Why did the largest four firms introduce a relatively large share of the innovations in some cases, but not in others? Consider the innovations of a particular type (that is, process or product) that were introduced during a given period of time in a particular industry. More particularly, consider those which required, for their introduction, a minimum investment of I.

Table 5.4. *Innovations and Capacity (or Output) Accounted for by Largest Four Firms, Steel, Petroleum-refining, and Bituminous Coal Industries, 1919–1938 and 1939–1958 (in percent)*

	PERCENT OF INDUSTRY TOTAL					
	STEEL[a]		PETROLEUM[b]		COAL[c]	
ITEM	WEIGHTED[d]	UN-WEIGHTED	WEIGHTED[d]	UN-WEIGHTED	WEIGHTED[d]	UN-WEIGHTED
1919–1938						
Process innovations	39	41	34	36	27	18
Product innovations	20	20	60	71	—	—
All innovations[e]	30	32	47	54	27	18
Capacity (or output)	62	62	33	33	11	11
1939–1958						
Process innovations	58	64	58	57	30	27
Product innovations	27	27	40	34	—	—
All innovations[e]	43	51	49	43	30	27
Capacity (or output)	63	63	39	39	13	13

SOURCE: Tables 5.1, 5.2, and 5.3, and Appendix B.

[a] Ingot capacity is used to measure each firm's size. The industry is defined to be those firms with ingot capacity, but firms with such capacity that are engaged primarily in some other business were excluded. For the earlier period, a firm's size refers to 1926; for the later period, it refers to 1945.

[b] Crude capacity is used to measure each firm's size. The industry is defined to be those firms with crude capacity. For 1919–1938, a firm's size refers to 1927; for 1939–1958, the figures refer to 1947. The product innovations included here are petrochemicals; in each case the innovator is the first petroleum company that produced it.

[c] Annual production is used to measure each firm's size. The industry is defined to include all that produced bituminous coal. For 1919–1938 a firm's size refers to 1933; for 1939–1958 these figures refer to 1953. The innovations included here are all new devices for preparing coal. This was the only kind of data I could obtain.

[d] In the columns headed "weighted," each innovation is weighted in proportion to its average rank by "importance" in the lists obtained. For processes I suggested that total savings be used to judge relative importance; for new products I suggested that sales volume be used.

[e] For the weighted data, this is just the unweighted average of the figures for process and product innovations.

Letting $\pi_j(I)$ be the proportion of these innovations introduced by the jth firm in this industry, we assume that

$$(1) \qquad \pi_j(I) = \begin{cases} 0 & \text{if } S'_j < M \\ B_1(I) + \beta_2(I)S'_j + E_j(I) & \text{if } S'_j \geq M, \end{cases}$$

where S'_j is the size (measured in terms of assets) of the jth firm. Of course, $B_1(I)$, $\beta_2(I)$, and M vary among industries, time periods, and types of innovations, $\beta_2(I)$ is presumed to be always positive, and $E_j(I)$ is a random error term.

Firms below a certain size (M) introduce none of the innovations because they lack the volume of production required to use the innovations profitably. For simplicity, we assume that the minimum size of a firm required to use an innovation is approximately the same for innovations of the same type that occur in a particular time interval in a given industry. For firms larger than M, we suppose that the proportion of these innovations introduced by a firm is a direct, linear function of its size, for the reasons discussed in section 1 of this chapter as well as those discussed in section 1 of Chapter 8.[9]

Next, we assume that a firm's size has more effect on $\pi_j(I)$ if the innovations require relatively large investments than if they can be introduced cheaply. If the innovations require very large investments, one would expect that larger firms would be required to finance these projects and to take the risks.[10] More specifically, we assume that

$$(2) \qquad \beta_2(I) = \alpha_1 + \alpha_2 I/\bar{S}_M + z,$$

where $\bar{S}_M$ is the average assets of the firms with assets greater than or equal to M, α_2 is presumed to be positive, and z is a random error term. The ratio of I to $\bar{S}_M$ (rather than I alone) is used because in the present context the size of the investment must be related to the average size of the relevant firms. Measures other than $\bar{S}_M$ (for example, M) could have been used instead, but the results would have been much the same as those presented below.[11]

[9] A linear function is only a convenient simplification. Up to some point, increases in size may bring progressively greater increases in $\pi_j(I)$ because a certain minimum size must be attained before a research laboratory can be maintained, if it is assumed that this size exceeds M. (See section 1 and R. Nelson, "Uncertainty, Learning, and the Economics of Parallel Research and Development Efforts," *Review of Economics and Statistics* [Nov. 1961].) Beyond some point, increases in size may result in less than proportionate increases in the number of innovations, because little further advantage is gained from the viewpoint of the pooling of risks or the ease with which innovations can be financed, and, as is often alleged, the motivation to innovate may become weaker and administrative difficulties may multiply. See Stocking, *Testimony* H. of R., 1950. Also, one might use a firm's percent of the industry's assets rather than its assets, in equation (1). See note 14. Finally, it has been suggested that another reason why $\pi_j(I)$ increases with firm size is that inventors may try first to get the largest firms interested in their inventions.

[10] The results of the interviews described in Chap. 8 seem to be consistent with this.

[11] Had I/M rather than $I/\bar{S}_M$ been used in equation (2), the results in equation (4) would have been

$$\pi - 4/N(M) = \underset{(0.00009)}{0.00013} [\bar{S}_4 - \bar{S}_M] + \underset{(0.0010)}{0.0032} I[\bar{S}_4 - \bar{S}_M]/M,$$

and the estimate of α_2 would still be positive and statistically significant.

If these assumptions hold, it follows that the proportion of all of the innovations (of a particular type that were introduced in a given period of time in a particular industry) carried out by the four largest firms should equal

$$\pi = 4/N(M) + 4\alpha_1[\bar{S}_4 - \bar{S}_M] + 4\alpha_2\bar{I}[\bar{S}_4 - \bar{S}_M]/\bar{S}_M + z', \quad (3)$$

where $N(M)$ is the number of firms with assets greater than or equal to M, $\bar{S}_4$ is the average assets of the four largest firms, $\bar{I}$ is the average minimum investment required to introduce these innovations, and z' is a random error term.[12] Thus, $N(M)$, $(\bar{S}_4 - \bar{S}_M)$, and $\bar{I}/\bar{S}_M$ determine whether or not the four largest firms introduce a disproportionately large share of the innovations.

According to this model, the characteristics (particularly $\bar{I}$ and M) of the innovations that can profitably be introduced in a particular industry during a given time interval are exogenous variables determined by the largely unpredictable nature of the technical breakthroughs made previously by members of the industry, equipment manufacturers, and independent research organizations. If, on the contrary, these characteristics are influenced by the extent to which the largest firms are the innovators, an identification problem arises in equation (3).

For example, if smaller (larger) firms in the industry, when confronted with various research and innovative opportunities, favor those with small (large) values of I and M, π may be directly related to $\bar{I}$ and M although the line of causation is the reverse of that underlying equation (3). Though an identification problem of this sort may turn out to be troublesome in some industries, interviews indicate that it is probably of little significance in the industries used here.[13] Lacking other evidence, we proceed on this assumption.

To see how well this model can explain the observed difference in π, we obtained rough estimates of $\bar{I}$, M, N(M), $\bar{S}_4$, and $\bar{S}_M$ for the innovations of each type in steel and petroleum during each period. Unfortunately, suitable data of this sort could not be obtained for coal. The results are shown in Table 5.5. Using these data, we derived least-squares estimates of α_1 and α_2.

[12] Since the sum of $\pi_j(I)$ is one, it follows that, if $f(S'_j)$ is the number of firms of size S'_j and if the sum of the $E_j(I)$ is zero,

$$1 = \sum_{S'_j > M} [B_1(I) + B_2(I)S'_j]f(S'_j),$$

$$= B_1(I)N(M) + B_2(I)N(M)\bar{S}_M$$

Thus, $$B_1(I) = [N(M)]^{-1} - B_2(I)\bar{S}_M.$$

Substituting this expression for $B_1(I)$ (and the expression for $B_2(I)$ in equation (2)) into equation (1), weighting $\pi_j(I)$ by the proportion of all innovations that require a minimum investment of I and summing over I to obtain the proportion of all innovations introduced by the jth firm, and summing the results for the four largest firms, we have equation (3). Of course, $z' = 4z[\bar{S}_4 - \bar{S}_M]$ plus the sum for the four largest firms of the average value of $E_j(I)$; and $B_2(I)$ must be greater than $N(M)[\bar{S}_M - M]^{-1}$. Using the estimates of $B_2(I)$, the latter inequality almost always seems to hold for $I = \bar{I}$.

[13] According to interviews with executives of engineering associations and research directors of firms, the line of causation has predominantly run in the direction presumed by the model. But such evidence is hardly conclusive, and the problem may be more serious than they indicate.

Table 5.5. *Values of M, N(M), $\bar{I}$, $\bar{S}_4$ and $\bar{S}_M$, Steel and Petroleum-refining Industries, Process and Product Innovations, 1919–1938 and 1939–1958*

INDUSTRY AND TYPE OF INNOVATION	PARAMETER[a]				
	M	N(M)	$\bar{I}$	$\bar{S}_4$	$\bar{S}_M$
			1919–1938		
Steel					
Process	46.0	19	0.60	858.5	245.2
Product	47.0	18	0.10	858.5	256.2
Petroleum					
Process	10.0	81	1.75	554.4	72.6
Product	18.2	50	3.30	554.4	109.2
			1939–1958		
Steel					
Process	26.3	29	1.30	1238.0	256.3
Product	23.5	30	0.50	1238.0	248.6
Petroleum					
Process	13.3	82	1.77	1243.0	144.0
Product	36.4	34	2.08	1243.0	314.7

SOURCE: See Appendix B.

[a] SYMBOLS: M, the average minimum size (assets) of firm required to use the innovations; N(M), the number of firms exceeding this size; $\bar{S}_M$, the average size (assets) of the firms exceeding this size; $\bar{S}_4$, the average size (assets) of the four largest firms; and $\bar{I}$, the average minimum investment required to install the innovations. All but N(M) are expressed in millions of dollars.

Inserting them into equation (3), we obtained

$$\text{(4)} \qquad \pi - 4/N(M) = \underset{(0.00007)}{0.00014}\,[\bar{S}_4 - \bar{S}_M] + \underset{(0.0063)}{0.0289}\,\bar{I}[\bar{S}_4 - \bar{S}_M]/\bar{S}_M,$$

where the figures in parentheses are standard errors and z' is omitted. As the model predicts, the estimate of α_2 is positive and statistically significant.[14]

[14] Note three things: First, because the error term would be more nearly homoscedastic, it might be argued that both sides of equation (3) should be divided by $(\bar{S}_4 - \hat{M}_M)$ and that least-squares estimates should then be made of α_1 and α_2. The resulting estimates, and their standard errors, turn out to be almost precisely the same as those in equation (4). Second, the model can be extended without too much difficulty to take account of differences among innovations in M. Third, if, as is suggested in note 9, $\pi_j(I)$ is assumed to be a linear function of $S'_j / \sum_{S'_j > M} S'_j$, rather than S'_j, it turns out that

$$\pi - \frac{4}{N(M)} \text{ should equal } V_1 \frac{\bar{S}_4 - \bar{S}_M}{N(M)\bar{S}_M} + V_2 \frac{\bar{I}}{\bar{S}_M} \frac{\bar{S}_4 - \bar{S}_M}{N(M)\bar{S}_M}.$$

An equation of this sort fits about as well as equation (4), the least-squares estimate of V_1 being 1.12 and that of V_2 being 183 (and statistically significant). It is difficult at this point to choose between these two forms of the model.

FIGURE 5.1. The plot of actual value of $\pi - 4/N(M)$ against that computed from equation (4), process and product innovations, steel and petroleum-refining industries, 1919–1958. The line is a 45-degree line through the origin. S = steel, P = petroleum, superscript ^{e} = earlier period, superscript ^{L} = later period, subscript $_r$ = process innovations, and subscript $_d$ = product innovations.

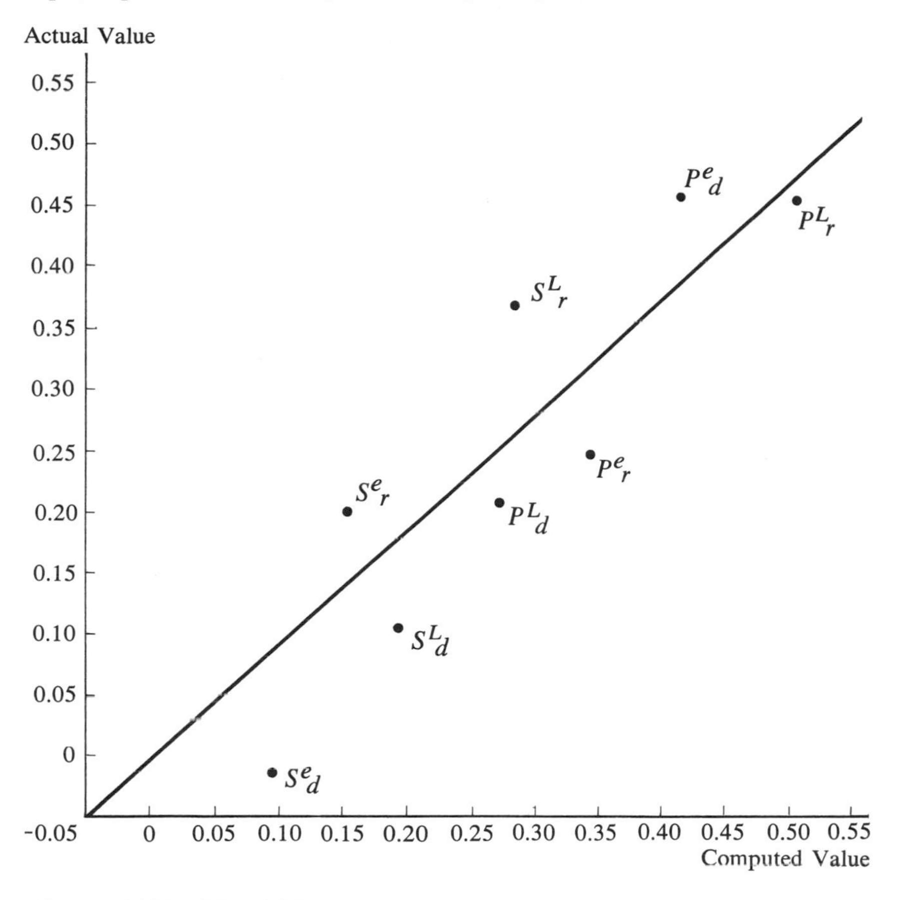

SOURCE: Tables 5.4 and 5.5.

Figure 5.1 shows that this equation can represent the data in Table 5.4 quite well, the coefficient of correlation (adjusted for degrees of freedom) being 0.88.[15] Moreover the results are rather insensitive to changes in Ī and M, the values of which (in Table 5.5) are probably subject to error. When

[15] Note two things: First, there is no tendency for the residuals from equation (4) to be positive in one industry or time period and negative in another. They seem quite random in this regard. Second, even if we include only firms larger than M, the largest four petroleum firms seem to account for a disproportionately large share of the product innovations in 1919–1938 and the process innovations in 1939–1958. For process innovations in 1919–1938 and product innovations in 1939–1958, they account for about the "expected" share. Of course, including only firms larger than M, the largest four steel firms fare even worse than in Table 5.4.

each of these estimates was varied by plus or minus 20 percent, the sign being randomly chosen, the results remained essentially the same (in all half-dozen trials) as in equation (4).

There is also some evidence that the model can predict the situation in the railroad industry reasonably well. Consider the extent to which the largest railroads have been the innovators. Healy includes the following in a list of ten major innovations introduced in the railroad industry since 1920: (1) four-wheel trailing trucks and large fire box for steam locomotives, (2) the diesel-electric switcher, (3) the diesel-electric passenger locomotive, (4) the diesel-electric freight locomotive, (5) the roller bearing, (6) the streamlined, light-weight passenger car, (7) the AB freight airbrake, (8) air conditioning, (9) car retarders, and (10) centralized traffic control. Taking this very small sample of innovations, we determined which firm was first to introduce each one and the size distribution of these innovators [92].

The results, shown in Table 5.6, seem to support the Schumpeterian view. Almost all of the innovators were among the ten largest firms, and about half were among the four largest. Moreover, although the largest four firms

Table 5.6. *Size Distribution of Innovators and All Class I Railroads*

SIZE OF FIRM (in millions of ton-miles, 1920)	FIRMS	NUMBER OF INNOVATIONS[a]	INNOVATORS
	Each System Regarded as a Unit		
Less than 4,000	77	0.0	0
4,001– 8,000	6	0.0	0
8,001–12,000	7	0.0	0
12,001–16,000	8	3.0	3
16,001–20,000	3	0.5	1
over 20,000	3	6.5	3
	Individual Class I Line-Haul Railroads[b]		
Less than 4,000	159	1.0	1
4,001– 8,000	9	0.0	0
8,001–12,000	12	2.0	2
12,001–16,000	4	1.0	1
16,001–20,000	1	0.5	1
over 20,000	3	4.5	3

SOURCE: See E. Mansfield, "Innovation and Technical Change in the Railroad Industry," *Transportation Economics*, National Bureau of Economic Research, 1965.

[a] When two firms began using an innovation at about the same time, the credit was split equally between them. Thus, fractions occur in this column.

[b] There is one less innovation in this part of the table than in the upper part, because one of the innovators, although part of a very large system, was not a Class I line-haul railroad.

account for about 32 percent of the industry's ton-miles, they account for 65 percent of the innovations. Thus, if one assumes that these innovations are a representative sample, the largest four firms accounted for a significantly larger share than would be expected on the basis of their share of the industry's ton-miles. By this criterion their share of the innovations was disproportionately large.

To what extent can equation (4), based on experience in the steel and petroleum industries, predict the behavior of the largest four firms in the railroad industry? Since the value of π derived from Table 5.6 could, with reasonable probability, depart from the true value by as much as 0.25, simply because of sampling errors, it is clear that a test of this sort will not be very powerful. Nonetheless, it is worth carrying out. Ignoring z', the estimate of π obtained from equation (4) is 0.54, which is reasonably close to 0.65, the value of π derived from Table 5.6. Thus, taking account of the independent variables in equation (4), it appears that, in comparison with other industries, the estimate of the largest four railroads' share of the innovations is higher than would be expected, but that the difference could easily be due to chance.

3. Dissolution of Corporate Giants

Thus far, we have considered whether or not the largest firms accounted for a disproportionately large share of the innovations. More basically, one would like to know whether fewer innovations would have been introduced during each period if these giants had been broken up. If one is willing to ignore the effects of *all* factors other than a firm's size on the number of innovations it carried out, some very rough answers can be obtained, but these results should obviously be treated with the utmost caution. It would be foolhardy to attach a great deal of policy significance to them.

We assume that

$$n_j = a_0 + a_1 \ln S_j + a_2(\ln S_j)^2 + a_3(\ln S_j)^3 + z'', \tag{5}$$

where n_j is the number of innovations carried out by the jth firm, S_j is the firm's size (measured in terms of physical capacity or output), and z'' is a random error term. This model is more useful and convenient than equation (1) for present purposes. Needless to say, the use of equation (5) rather than equation (1) involves no more than a substitution of one sort of approximation for another.[16] The logarithm of S_j is used, rather than S_j, because its distribution is less skewed.

Let the computed regression in equation (5) be $N(S_j)$ and ignore the

[16] As we pointed out in note 9, equation (1) ignored the fact that n_j might be a curvilinear function of S'_j; on the other hand, equation (5) ignores the fact that n_j may be zero below some value of S_j. Equation (5) is more convenient here because ordinary regression techniques can be used to estimate the a's, and its disadvantages are avoided in large part by excluding the very smallest firms.

sampling errors in it. Then, if $N(S_j)/S_j$ is a maximum at or very near the size of the largest firm, it would appear on the basis of these assumptions that the dissolution of the largest firms would have resulted in fewer innovations being carried out. On the other hand, if $N(S_j)/S_j$ reaches a maximum far below the size of the largest firm, their dissolution would presumably have had a positive effect.[17]

Such an analysis is extremely crude, but what do the results suggest? Table 5.7 contains the estimates of the a's, the great majority of which are highly significant in a statistical sense and (taking account of differences in the total number of innovations) reasonably stable over time. In the petroleum and coal industries, the regressions indicate that $N(S_j)/S_j$ reaches a maximum at about the size of the sixth largest firm. Thus, they suggest that fewer innovations would have been introduced if firms other than the top few had been broken up. In view of the sampling errors in the regressions, the true maximum value of $N(S_j)/S_j$ may occur at an even larger value of S_j, and the dissolution of the largest few firms might also have had deleterious effects.[18]

On the other hand, the results in the steel industry are in complete conflict with the Schumpeterian view, the maximum value of $N(S_j)/S_j$ always being found among very small firms. These findings lend support to Stocking's assertion [161] that the largest steel producers have not tended to be the technical leaders. Judging by our results, it would be difficult to justify the size of these firms on the basis of their past performance as innovators.[19]

Finally, several additional points should be noted regarding Table 5.7. First, a firm's residual from a process-innovation regression is positively but weakly correlated in each case with its residual from the corresponding product-innovation regression, the correlation coefficient being about 0.3. Thus, when the size of the firm was held constant, a firm that did considerable innovating with regard to processes also tended to do considerable innovating with regard to products. Second, if S_j is used rather than $ln\ S_j$ in equation (5), the results are generally like those described above. The maximum value of

[17] See Appendix B for some of the difficulties in this procedure. Note that the results of previous chapters do not indicate that the dissolution of corporate giants would result in a smaller number of innovative opportunities—i.e., inventions—in most industries for which we have data.

[18] If $N(S_j)/S_j$ is plotted as a function of S_j, one finds that its maximum occurs at about 200,000 barrels of crude capacity (1919–1938), 300,000 barrels of crude capacity (1939–1958), 3,600,000 tons of coal (1919–1938), and 7,800,000 tons of coal (1939–1958). In the steel industry, the maximum occurs in every case at less than 1,000,000 tons of ingot capacity.

[19] Note two points. First, the future situation in steel may be quite different. In the interviews described in Appendix B, several executives claimed that U.S. Steel was becoming much more of an innovator than in the past. Second, although U.S. Steel has frequently been criticized on this score, its performance seems to be much better than the second-largest firm—Bethlehem.

For discussions of the situation in petroleum, see J. Bain, *The Economics of the Pacific Coast Petroleum Industry* (University of California, 1944), Vol. 3. For both steel and petroleum, see D. Hamberg, *Testimony on Employment, Growth, and Price Levels Before the Joint Economic Committee of Congress*, 1959.

Table 5.7. *Least-squares Estimates of a_0, a_1, a_2, and a_3, Steel, Petroleum-refining, and Bituminous Coal Industries, Process and Product Innovations, 1919–1938 and 1939–1958*

INDUSTRY AND TYPE OF INNOVATION[a]	a_0	a_1	a_2	a_3	CORRELATION COEFFICIENT	NUMBER OF FIRMS
			1919–1938			
Steel						
Process	−1.12	0.901[b]	−0.221[b]	0.017[b]	0.73	101
		(.323)	(.063)	(.004)		
Product	−0.62	.583	−.144[b]	.011[b]	.43	101
		(.324)	(.063)	(.004)		
Petroleum						
Process	−0.02	.135[b]	−.103[b]	.021[b]	.64	269
		(.052)	(.027)	(.004)		
Product	−0.04	.305[b]	−.245[b]	.045[b]	.78	269
		(.060)	(.031)	(.004)		
Coal						
Process	0.04	.031[b]	.024[b]	.009[b]	.28	639
		(.011)	(.007)	(.004)		
			1939–1958			
Steel						
Process	−0.82	.624	−.153	.012[b]	.72	68
		(.662)	(.117)	(.006)		
Product	0.59	.331	.048	−.001	.58	68
		(.434)	(.077)	(.004)		
Petroleum						
Process	−0.05	.235[b]	−.155[b]	.025[b]	.83	269
		(.040)	(.018)	(.002)		
Product	−0.07	.300[b]	−.208[b]	.035[b]	.70	269
		(.010)	(.043)	(.005)		
Coal						
Process	0.02	.038[b]	.046[b]	.015[b]	.51	582
		(.014)	(.004)	(.003)		

SOURCE: Tables 5.1, 5.2, and 5.3.

[a] The smallest firms were omitted. See note 20.

[b] Significant at 0.05 probability level.

$N(S_j)/S_j$ is found among the very largest firms in the petroleum and coal industries and among very small firms in the steel industry.[20]

[20] To what extent were the significant innovations in these industries introduced by new firms? According to some authors, society frequently must rely on such firms to be the innovators. For example, see E. Domar, "Investment, Losses, and Monopoly," *Income, Employment, and Public Policy* (New York: Norton, 1948); and Mansfield, *Economics of*

Third, it is possible of course that the average delay in utilizing inventions, not the number utilized, would be affected if the largest firms were broken up. Section 5 investigates this possibility in the case in which the inventor is not a member of the industry. Fourth, judging by the correlation coefficients in Table 5.7, equation (5) fits the data only moderately well. For the innovations in coal and the product innovations in steel, the fit is particularly unimpressive.

Fifth, the regressions indicate that in all cases n_j is an increasing function of S_j throughout all, or almost all, of the relevant range—which is what one would expect. Sixth, the data on which the regressions are based are quite consistent with the estimates of M in Table 5.5. Firms smaller than M seldom, if ever, introduced an innovation in these industries.[21]

Finally, we have assumed that all innovations can be lumped together. If we were to separate innovations into various categories, depending on their characteristics and functions, the estimates of the a's undoubtedly would vary from one category to another and $N(S_j)/S_j$ would reach a maximum at a different value of S_j in one category than in another. If it is optimal for innovations in various categories to be carried out, a considerable diversity of firm sizes would probably be socially desirable; certainly this seems reasonable. With more detailed data, it would not be difficult to extend the analysis in this direction.

4. The Changing Role of Large and Small Firms

It is frequently asserted that a small firm does less innovating now, relative to a large firm, than it did in the past.[22] Because of rising development costs and the greater complexity of technology, this hypothesis seems plausible for a wide range of industries. Does it hold in steel, petroleum, and coal?

To help answer this question, we use the regressions in equation (5) to compute the ratio, for each value of S_j, of the average value of n_j in 1939–1958

Technological Change. But in industries like steel and petroleum, where entry was difficult and the industry was mature, this seems less likely. As it turns out, not one of the innovations for which we have data was introduced by a new firm. In the coal industry, the results are the same—but the data may be biased somewhat. The innovations in the coal industry are new techniques for preparing coal. According to interviews with the Bureau of Mines personnel, such techniques were probably less likely to be introduced by new firms and small firms than many other types. However, this bias is unlikely to be great enough to reverse the results in this section. Had it been possible to include other types of innovations as well, the largest firms would probably have continued to account for a disproportionately large share, but the difference might not have been so large. Note that the equipment producers had a very important hand in developing most of these innovations.

[21] Of course, the fact that a few firms below our estimate of M are innovators does not mean that our estimates are incorrect. We assume in section 2 that M is the same for all innovations, but this really is not the case and our estimates are really of the average value of M. Thus, the size of some innovators may fall below them. Note too that a variant of the Tobin-Rosett technique for handling limited dependent variables might have been used to estimate M.

[22] For example, see Hamberg, *Testimony on Employment*, 1959, for some discussion of this.

to its average value in 1919–1938. That is, taking a particular industry and a particular type of innovation and letting $\hat{a}_0 \ldots \hat{a}_3$ stand for the 1939–1958 estimates of the a's and $\hat{a}'_0 \ldots \hat{a}'_3$ stand for the 1919–1938 estimates of the a's, we compute

$$[\hat{a}_0 + \hat{a}_1 \ln S_j + \hat{a}_2(\ln S_j)^2 + \hat{a}_3(\ln S_j)^3] / [\hat{a}'_0 + \hat{a}'_1 \ln S_j + \hat{a}'_2(\ln S_j)^2 + \hat{a}'_3(\ln S_j)^3]$$

for each value of S_j. If this ratio increases as S_j increases, it suggests that the small firms have become less important, on the average, as a source of innovations, relative to large firms.

Table 5.8. *Average Number of Innovations Carried Out by Firms of Given Size in 1939–1958 Divided by Average Number Carried Out by Firms of the Same Size in 1919–1938, Steel, Petroleum-refining, and Bituminous Coal Industries*

	RATIO OF AVERAGE NUMBER OF INNOVATIONS IN 1939–1958 TO AVERAGE NUMBER IN 1919–1938	
SIZE OF FIRM	PROCESS	PRODUCT
Steel (ingot capacity, tons)		
100,000	0.22	0.00[a]
1,000,000	0.93	6.50
5,000,000	0.84	1.25
10,000,000	0.80	0.90
25,000,000	0.78	0.65
Petroleum (crude capacity, barrels)		
30,000	0.00[a]	—[b]
60,000	0.00[a]	0.45
120,000	0.47	0.60
240,000	0.66	0.65
480,000	0.77	0.68
Coal (production, tons)		
600,000	0.44	—
1,200,000	0.68	—
2,400,000	1.14	—
4,800,000	1.43	—
9,600,000	1.59	—

SOURCE: Table 5.7.

[a] According to the regression in equation (5), the average number of innovations in 1939–1958 was negative. It seems reasonable in this context to substitute zero for such negative numbers.

[b] According to the regressions in equation (5), both the average number of innovations in 1919–1938 and that in 1939–1958 were negative. Thus, the ratio would have no meaning.

The results indicate that in all three industries and for both types of innovations, the ratio was higher among the largest firms than among the smallest ones. (See Table 5.8 for the ratio at selected values of S_j.) Thus, the smallest firms have in each case become less important in this respect.[23] The estimates in Table 5.5 shed some additional light on the factors causing this decline in the relative importance of the smallest firms. In steel, it may have been due to an increase in the capital requirements for innovating, but not to an increase in the minimum size of firm that could profitably use the innovations. In petroleum, it may have been due to an increase in the minimum size of firm that could profitably use the innovations, but not to an increase in capital requirements. Unfortunately, estimates of M and $\bar{I}$ are not available for the bituminous coal industry.

Finally, one additional point should be noted. Whereas the ratio of the average value of n_j during 1939–1958 to its average value in 1919–1938 rises monotonically with S_j in the petroleum and coal industries, it rises and then falls in steel. Thus, relative to a very large firm, a medium-sized steel firm introduced, on the average, a larger number of innovations during 1939–1958 than during 1919–1938.[24] The reasons for this are by no means clear, but it may have been that the medium-sized firms increased their R and D activities by a greater proportion than did the large firms. This would be consistent with a hypothesis put forth by Schmookler [144].[25]

5. Market Structure and the Rate of Introduction of Inventions

Suppose that an individual or firm invents a device that could be used profitably in a particular industry, but suppose also that the inventor is not a member of this industry and that consequently he must induce some firm in the industry to introduce it or enter the industry himself. For this type of invention, an important question is: What effect would a change in market structure have on the length of time that elapses before someone introduces the invention? This question has received considerable attention—both recently and in the past. On the one hand, there are some—like Bain [10], Brozen [18], Joan Robinson [133], and Stocking [161]—who believe that

[23] These results are very rough. For one thing, they assume that the 1919–1938 relationship between n_j and S_j can be extrapolated somewhat, since the largest firms in 1939–1958 were larger than those in 1919–1938.

[24] The range of firm sizes in Table 5.8 was chosen so that it just about covers the range in which the innovators lie. Thus, very small steel firms must be included because they have done a good deal of the innovating.

[25] He asserts that "given the progressive improvement in the quality of management . . . given the growing recognition . . . of the value of research, and given the increasing supply of engineers and scientists, a rise in the relative importance of organized research and development among small and medium-sized firms is perhaps to be expected." (See Schmookler, "Bigness, Fewness, and Research," p. 631). Our results could be due in part to such a movement in the past, since it would be expected that the medium-sized firms would react before the small ones. The relevant results in Chapter II do not seem to bear this out, but they refer to a later period and to only the largest firms.

inventions would be applied most rapidly under purely competitive conditions. They argue that if many firms exist, there is more protection against an invention's being blocked by the faulty judgment of only a few men. Moreover, they allege that the existence of many competitors will force a firm to seek out and apply new ideas, whereas a live-and-let-live policy may develop otherwise.

On the other hand, there are others—like Villard [167]—who think that they would be applied most rapidly if industries contained relatively few firms but large ones. They point out that such firms are better able to finance the introduction of inventions and to take the necessary risks. In addition, they claim that the larger firms will have better managers who will be more inclined to innovate. Although each group has some convincing points to make, there is no evidence that one's arguments are universally more powerful than the other's. And in a particular case we are unable to tell how these factors should be quantified and weighted so that a conclusion can be reached. This section contains some exploratory attempts to devise operational techniques to handle this problem.

Suppose that an industry is composed of $n - 1$ firms. If at time t a particular invention of this sort has not yet been introduced, suppose that the probability that the ith firm will introduce it between time t and time $t + \Delta$ is $\lambda_i\Delta$. Suppose that the probability that the inventor or some other new entrant into the industry will introduce it then is $\lambda_n\Delta$. Assume too that there is no collusion among the firms to prevent the application of the technique. That is, assume that the probability that the ith firm will introduce it between time t and time $t + \Delta$ does not depend on whether some other firm decides to do so.[26]

Under these conditions, one can easily obtain expressions for L, the expected length of time that will elapse before the invention is applied, and P_j, the probability that the jth firm will be the innovator. Letting Δ become very small so that we approach a continuous measurement of time, it can easily be seen that

$$L = \lim_{\Delta \to 0} \Delta \sum_{r=1}^{\infty} rp(r),$$

where $p(r)$ is the probability that it will take r periods of length Δ before the invention is applied. Letting θ be the probability that no existing or new firm introduces it during a period of length Δ and $(1 - \theta)$ be the probability that one or more existing or new firms introduces it during such a period, it follows that

[26] All this is assumed to hold only for small Δ. In the analysis below, Δ tends to zero and terms of higher order in Δ would vanish. We assume that the development work has already been done by the inventor or that it will take about the same length of time regardless of which firm does it. Of course this may not be the case. Attention is focused strictly on the mean delay, but for some purposes higher moments might also be relevant. Finally, the arguments presented at the beginning of this section are only a few of those that have been used.

$$
\begin{aligned}
L &= \lim_{\Delta\to 0} \Delta \sum_{r=1}^{\infty} r(1-\theta)\theta^{r-1}, \\
&= \lim_{\Delta\to 0} \Delta(1-\theta) \sum_{r=1}^{\infty} r\theta^{r-1}, \\
&= \lim_{\Delta\to 0} \Delta(1-\theta) \left[\frac{d}{d\theta} \sum_{r=0}^{\infty} \theta^{r}\right], \\
&= \lim_{\Delta\to 0} (1-\theta) \left[\frac{d}{d\theta} \frac{1}{(1-\theta)}\right] \\
&= \lim_{\Delta\to 0} \Delta/(1-\theta)
\end{aligned}
$$

Since $\theta = \prod_{j=1}^{n} (1-\lambda_j\Delta)$,

$$
\begin{aligned}
L &= \lim_{\Delta\to 0} \frac{\Delta}{1 - \prod_{j=1}^{n} (1-\lambda_j\Delta)}, \\
&= \lim_{\Delta\to 0} \frac{\Delta}{1 - 1 + \sum_{j=1}^{n} \lambda_j\Delta + o(\Delta)},
\end{aligned}
$$

where $o(\Delta)$ stands for terms which tend to zero when, after dividing by Δ, we let Δ tend to zero. Consequently,

$$
L = \lim_{\Delta\to 0} \frac{1}{\sum_{j=1}^{n} \lambda_j + \frac{o(\Delta)}{\Delta}}
$$

$$
L = \left[\sum_{j=1}^{n} \lambda_j\right]^{-1}. \tag{6}
$$

Similarly, it is apparent that

$$
P_i = \lim_{\Delta\to 0} \sum_{r=1}^{\infty} G_i(r),
$$

where $G_i(r)$ is the probability that the ith firm will introduce the invention in the rth period and that it will be the first firm to do so. Letting ρ_i be the probability that only the ith firm will introduce it during the rth period,

$$
\begin{aligned}
P_i &= \lim_{\Delta\to 0} \sum_{r=1}^{\infty} \rho_i\theta^{r-1} \\
&= \lim_{\Delta\to 0} \rho_i \sum_{r=1}^{\infty} \theta^{r-1} \\
&= \lim_{\Delta\to 0} \rho_i/(1-\theta).
\end{aligned}
$$

Since $\rho_i = \lambda_i\Delta \prod_{\substack{j=1 \\ \neq i}}^{n} (1 - \lambda_j\Delta)$,

$$P_i = \lim_{\Delta \to 0} \left[\frac{\lambda_i\Delta \prod_{\substack{j=1 \\ \neq i}}^{n} (1 - \lambda_j\Delta)}{1 - \prod_{j=1}^{n} (1 - \lambda_j\Delta)} \right],$$

$$= \lim_{\Delta \to 0} \left[\frac{\lambda_i\Delta(1 + u(\Delta))}{1 - 1 + \sum_{j=1}^{n} \lambda_j\Delta + o(\Delta)} \right]$$

$$= \lim_{\Delta \to 0} \left[\frac{\lambda_i(1 + u(\Delta))}{\sum_{j=1}^{n} \lambda_j + \frac{o(\Delta)}{\Delta}} \right],$$

where $u(\Delta)$ stands for terms that vanish as Δ tends to zero. Thus

$$P_i = \lambda_i / \sum_{j=1}^{n} \lambda_j$$

$$P_i = \lambda_i L. \tag{7}$$

Suppose that a change is being contemplated in the size distribution of firms in a given industry. Assume that each firm's size and its value of λ have been relatively constant in the recent past, and that once this reorganization is either carried out or dropped each firm's size and its value of λ will again remain relatively constant for some time.[27] Suppose that if the proposed reorganization occurs the frequency distribution of firms by size will be $n(S)$; if it does not occur suppose that it will be $m(S)$. Ignore differences among inventions in a firm's value of λ and the possibility that the inventor will enter the industry. Although they complicate things, these matters can be introduced without altering the essentials of the argument.[28]

[27] This simplification can be relaxed. All that we need to assume is that the future can be divided into epochs within which each firm's size and value of λ is relatively constant, that these epochs are very long relative to L, and that forecasts are available of the size distribution of firms (given that the reorganization does or does not occur) in each epoch. Then one can estimate the effects in each epoch.

[28] These factors can be introduced in the following way: The relationship between a firm's size and its value of λ is likely to differ, depending on the characteristics of the innovation. Thus, one should classify innovations by their capital requirement and other characteristics causing differences in the shape of this relationship. Factors (like the over-all profitability of the invention) that cause all the λ's to increase or decrease in proportion may be ignored. (If all λ_i's vary in proportion, the P_i will remain constant, and $\hat{f}(S)$ and the result in equation (9) will be unaffected. Inventions with proportional λ_i can be lumped into one class so long as the composition of the class with regard to profitability, etc., is unlikely to change much over time.) Classes should be established so as to maximize differences—ignoring proportional variations in all λ_i—in the shape of the relationships. Then $\hat{f}(S)$ can be estimated in each class, the goodness of fit being some indication of how homogeneous the class is. In each class, equation (9) can be used to estimate the percentage reduction in delay. To obtain an over-all estimate of the effect on all innovations, one must forecast the proportion of the

What effect will the proposed reorganization have on L? It seems reasonable to believe that a firm's value of λ is a function of its size. Suppose that whether or not the proposed change in market structure occurs the average value of λ for firms of given size will be proportional to the average value in the recent past (the coefficient of proportionality being ϕ). Of course, whether or not this is true depends on the particular change in market structure and on the characteristics of the industry.[29]

If this assumption holds and if $\lambda(S)$ was the average value of λ for firms of size S in the recent past, the expected delay, given that the reorganization occurs, is

$$L_o = [\phi \sum_s \lambda(S) n(S)]^{-1}. \tag{8}$$

And if $\hat{f}(S)$ is the regression in the recent past of the proportion of the innovations of this type that a firm carried out on its size, an estimate of the percentage change in average delay resulting from the proposed change in the size distribution of firms is

$$C = 100\{[\sum_s m(S)\hat{f}(S) / \sum_s n(S)\hat{f}(S)] - 1\}. \tag{9}$$

Since $n(S)$ and $m(S)$ are given and $\hat{f}(S)$ can be estimated from past data, the expression in equation (9) can be evaluated.[30]

As an illustration, consider process innovations in steel. In 1926, suppose that we had wanted to estimate the effect of splitting U.S. Steel into four smaller firms of equal size and keeping the rest of the steel producers at their 1926 size, the alternative being that all firms (including U.S. Steel) would maintain their 1926 size. Assuming that λ_i did not vary much from

---

innovations in the period ahead that will be in each class and estimate the average delay in each class in the previous period. These data—and the classes—must often be rough, but it is difficult to see how any technique could be devised that would not require them. If λ_n in the period ahead remains in the same proportion to the average value of λ in each size class (whether or not the reorganization occurs) one can easily handle the possibility of new entrants being the innovators. The proportion of the innovations carried out by new entrants is an estimate of P_n in the past—which equals $\phi\lambda'm$, where $\lambda'm$ is the value of λ_n in the future. Hence, this proportion can merely be added to both the numerator and denominator of the first term in parentheses on the right-hand side of equation (9). Essentially, this assumes that the reorganization will not seriously impede or promote entry into the industry. Whether or not this is true depends on the particular reorganization.

[29] Can the facts in section 3 be brought to bear on this assumption? Suppose that all the innovations included there were of the type considered here and that λ_i did not vary much from invention to invention. According to equation (7), λ_i is proportional to P_i, and the proportion of innovations carried out by the ith firm is an estimate of P_i. Take firms that resulted from important mergers in 1925–1938 and compare their proportion of the innovations in 1939–1958 with those of other firms of their (new) size. If a firm's value of λ_i adjusts relatively quickly to a change in its size, as we assume, their proportion should not differ significantly from the others. In fact, this turns out to be the case. This is an interesting but crude test of this assumption.

[30] Note that this regression should only include data for innovations invented outside the industry.

invention to invention and that the relation between the proportion of the innovations of the relevant kind that a firm carried out in the period immediately before 1926 and its 1926 size was like that between the proportion of all process innovations it carried out during 1919–1938 and its 1926 size,

$$\hat{f}(S) = -0.09 + 0.075 \ln S - 0.018(\ln S)^2 + 0.001(\ln S)^3. \quad (10)$$

To derive equation (10), merely divide equation (5)—after inserting the estimates of the a's in Table 5.7 into it—by the number of process innovations during 1919–1938, which is twelve.

Since $S = 22{,}628$ for U.S. Steel, it follows from equations (9) and (10) that the average delay, according to these crude estimates, would have been reduced by about 15 percent if U.S. Steel had been broken up in this way. To derive this figure, note that $\sum_s m(S)\hat{f}(S)$ is necessarily one because the residuals from a least-squares regression sum to zero. Moreover, $\sum_s n(S)\hat{f}(S)$ equals $1 - \hat{f}(22{,}628) + 4\hat{f}(5{,}657)$, since all firms other than U.S. Steel will maintain their size, even if the reorganization occurs. Since equation (10) shows that $\hat{f}(22{,}628) = 0.26$ and $\hat{f}(5{,}657) = 0.11$, it follows that $C = -15$. In absolute terms, a 15 percent reduction in delay is likely to be a reduction of at least 1 year.

Needless to say, these results can only be suggestive, since a change of this magnitude in an industry's structure might alter the relationship between the average value of λ and firm size.[31] This technique is likely to be more useful when the changes in market structure are less drastic.

6. Summary and Conclusion

The principal conclusions of this chapter are as follows: First, it is often alleged that the largest firms introduce a disproportionately large share of the innovations. For the first time, data of more than a fragmentary nature have been collected to test this hypothesis in a small number of important industries. The results vary from one industry to another. In petroleum refining, bituminous coal, and railroads, the largest four firms accounted for a larger share of the innovations than they did of the market. But in steel they accounted for less.

Second, the largest four firms seemed to account for a relatively large share of the innovating in cases in which (1) the investment required to innovate was large relative to the size of the potential users, (2) the minimum size of firm required to use the innovations profitably was relatively large, and (3) the average size of the largest four firms was much greater than the average size of all potential users of the innovations. A simple model that

[31] In addition, the assumptions underlying equation (10) are obviously very rough. For example, a firm's growth rate and other characteristics may affect its value of λ. We assume that these effects can be represented by the usual sort of random error term. Although this may not be too bad an approximation, attempts should be made in future research to include additional independent variables in equations (5) and (10).

focused particular attention on these factors could explain most of the observed interindustry and temporal differences.

Third, some very rough estimates suggest that in the petroleum and bituminous coal industries during 1919–1958 the largest few firms carried out no more innovations, relative to their size, than did somewhat smaller firms. In the steel industry, they carried out no more than considerably smaller firms. These estimates are crude and should be treated with the utmost caution.

Fourth, there is evidence that the smallest steel, oil, and coal firms did less innovating—relative to large and medium-sized firms—in recent years than in the period before World War II. In steel, this may have been due to an increase in the average investment required to innovate; in petroleum, it may have been due to an increase in the minimum size of firm that could use the innovations profitably.

Fifth, under certain circumstances, one can estimate the effect of a proposed change in market structure on the average time interval that elapses before an invention made outside the industry is applied. If a simple model of the innovation process holds, historical data identifying the innovators can be used to estimate these effects. Obviously, however, this technique has very important limitations and is only a first step toward the general solution of this problem.

CHAPTER 6

Innovation: Its Timing and Effects on Investment and Firm Growth

Economists have long been interested in the timing of innovation and its effects on investment and firm growth. In this chapter, we investigate these aspects of the innovative process. First, we consider the lag between invention and innovation. Second, data concerning the first commercial introduction of about 150 new processes and products in the iron and steel, petroleum-refining, and bituminous coal industries are used to determine whether the rate of occurrence of major innovations varies appreciably over the business cycle and, if so, how it varies.

Third, we formulate and test a simple investment function that includes the timing of innovation as an exogenous variable. This factor, which is ignored in practically all other econometric models, is clearly an important determinant of the level of investment. Fourth, estimates are made of the difference in growth rate between firms that carried out significant innovations and other firms of comparable initial size. The results help to measure the importance of successful innovation as a cause of interfirm differences in growth rates, and they shed new light on the rewards for such innovations.

1. The Lag Between Invention and Innovation

How long is the lag between invention and innovation? This lag must vary substantially, since some inventions require changes in tastes, technology, and factor prices before they can profitably be utilized, whereas others do not. Moreover, some inventions constitute major departures from existing practice and open up entirely new branches of technology, whereas others are more routine "improvement" inventions. In addition, in some cases the inventor and innovator are the same firm, whereas in other cases they are not. If we restrict our attention to relatively important inventions, the only data are extremely rough, since they are not based on random samples, and such concepts as "invention" and "innovation" are not easy to pinpoint and date.

Nonetheless, the available data provide some feel for the distribution of the lag.

Enos [34] estimated the time interval between invention and innovation for eleven important petroleum-refining processes and thirty-five important products and processes in a variety of other industries. Table 6.1 shows that it

Table 6.1. *Estimated Time Interval Between Invention and Innovation, Forty-Six Inventions, Selected Industries*

INVENTION[a]	INTERVAL (years)	INVENTION[a]	INTERVAL (years)
Distillation of hydrocarbons with heat and pressure (Burton)	24	Steam engine (Watt)	11
Distillation of gas oil with heat and pressure (Burton)	3	Ball-point pen	6
Continuous cracking (Holmes-Manley)	11	DDT	3
Continuous cracking (Dubbs)	13	Electric precipitation	25
"Clean circulation" (Dubbs)	3	Freon refrigerants	1
Tube and Tank process	13	Gyro-compass	56
Cross process	5	Hardening of fats	8
Houdry catalytic cracking	9	Jet engine	14
Fluid catalytic cracking	13	Turbo-jet engine	10
Catalytic cracking (moving bed)	8	Long playing record	3
Gas lift for catalyst pellets	13	Magnetic recording	5
Safety razor	9	Plexiglas, lucite	3
Fluorescent lamp	79	Nylon	11
Television	22	Cotton picker	53
Wireless telegraph	8	Crease-resistant fabrics	14
Wireless telephone	8	Power steering	6
Triode vacuum tube	7	Radar	13
Radio (oscillator)	8	Self-winding watch	6
Spinning jenny	5	Shell molding	3
Spinning machine (water frame)	6	Streptomycin	5
Spinning mule	4	Terylene, dacron	12
Steam engine (Newcomen)	6	Titanium reduction	7
		Xerography	13
		Zipper	27

SOURCE: J. Enos, "Invention and Innovation in the Petroleum Refining Industry," *The Rate and Direction of Inventive Activity* (Princeton, 1962), pp. 307–308.

[a] The first eleven inventions were those that occurred in petroleum refining.

averaged about 11 years in the petroleum industry and about 14 years in the others. Its standard deviation is about 5 years in the petroleum industry and 16 years in the others.[1] Enos [34] concludes that: "Mechanical innovations appear to require the shortest time interval, with chemical and pharmaceutical

[1] Of course, one would expect the standard deviation to be smaller for innovations in one industry than for those in many industries combined. Interindustry differences are included in the latter case.

innovations next. Electronic innovation took the most time. The interval appears shorter when the inventor himself attempts to innovate than when he is content merely to reveal the general concept.[2]

Lynn [74] estimated the average number of years elapsing from the basic discovery and establishment of an invention's technical feasibility to the beginning of its commercial development, as well as the average number of years elapsing from the beginning of the commercial development to its introduction as a commercial product or process. The results, based on twenty major innovations during 1885–1950, seem to suggest that the lag has been decreasing over time, that it is shorter for consumer products than industrial products, and that it is shorter for those developed with government funds than for those developed with private funds.

2. The Timing of Innovation

The literature abounds with theories regarding the way in which innovations are distributed over the business cycle. According to Graue [44], "a commonly accepted view is that inventions are introduced when industry is operating at low levels and when the pressure of competition is most unrelenting. . . ."[3] In their studies of new types of equipment, Brown [17] and Mack [77] seem to share this view. They argue that new designs tend to be postponed during good times and that ideas that accumulate are tried out, and new ones explored, during recessions.[4]

On the other hand, Keirstead [64] and Carter and Williams [22] believe that a strong seller's market is most favorable to the introduction of new processes and products. According to Keirstead, when the economy is approaching the top of the cycle, entrepreneurs "will welcome cost–reducing or output–expanding inventions. But when the economy turns down, . . . entrepreneurs will be chary of taking up inventions, except such as offer great cost reductions at a small increase in fixed costs."[5]

Almost no evidence exists on this score. The Conference on Price Research [26], having brought together a variety of statistical bits and pieces, concluded that the data "suggest a cyclical pattern in which the most rapid rate of technical change occurs during the middle of the expansion phase, with a slackening off toward the peak, some increase in the rate in the earlier part

[2] J. Enos, "Invention and Innovation in the Petroleum Refining Industry," *The Rate and Direction of Inventive Activity* (Princeton, 1962), p. 309.

[3] E. Graue, "Inventions and Production," *Review of Economics and Statistics* (Nov. 1953), p. 222.

[4] For some criticism of Brown's paper, see B. Perles, "Innovation in the Machine Tool Industry: Comment," *Quarterly Journal of Economics* (Aug. 1963). In recent years, there has been somewhat less talk about business cycles or fluctuations, due in part to the relatively uninterrupted growth of the economy during the 60's.

[5] B. Keirstead, *The Theory of Economic Change* (New York: Macmillan, 1948), p. 145. Note that J. Schmookler has argued that innovations occur after the crest of the boom ("Invention, Innovation, and Business Cycles," *Some Elements Shaping Investment Decisions* [Joint Economic Committee of Congress, 1962]).

Table 6.2. *Dates of First Commercial Introduction of Process and Product Innovations, Iron and Steel Industry, 1919–1958*

PROCESSES		PRODUCTS	
INNOVATION	DATE	INNOVATION	DATE
Continuous butt-weld pipe	1923	Valve steels	1919
Continuous wide-strip mill	1924	Nickel-bearing electrical steel	1920
Dolomite gun	1925	18-8 stainless steel	1926
Continuous pickling	1926	Nitriding steels	1927
Coreless induction electric furnace	1928	High-temperature alloys	1930
Automatic operation of open hearth	1928	Nonaging steel	1931
Electrically welded pipe	1929	High-strength low-alloy steels	1933
Multiple block wire drawing machines	1932	5 percent chrome hot work tool steel	1935
Electronic inspection of tin plate	1934	Grain-oriented electrical steel	1935
Austempering	1936	Low-tungsten high-speed tool steel	1935
Continuous annealing	1936	Boron-treated steels	1936
Continuous galvanizing	1936	Killed Bessemer steel	1939
Mechanical scarfing	1936	Aluminum-clad sheets	1939
Electrolytic tin plate[1]	1936	Manganese stainless steel	1943
Electric eye for Bessemer turn-down	1939	Titanium-treated enameling steel	1944
Stretch process for hot-reducing tube	1940	Extra-low-carbon stainless steel	1950
Sendzimir cold mill	1941	Precipitation hardening stainless steel	1955
Ultrasonic testing	1945	Columbium treated high-strength steel	1958
High top-pressure blast furnace	1945		
Carbon lining for blast furnace	1945		
Oxygen in melting	1946		
All-basic open hearth	1947		
Closed television circuits	1949		
Jet tapper	1949		
Continuous casting	1949		
Hot extrusion	1951		
Differential coating of tin plate[1]	1952		
Vacuum melting	1954		
L-D oxygen process	1954		
Vacuum degassing	1956		
Automatic programing of mills	1956		

SOURCE: See section 3.

[1] A case could be made for including these innovations as product, not process, innovations.

of the contraction, and another slackening off toward the trough." [6] In contrast, Healy [52] concluded that in the railroad industry there was no strong tendency for innovations to occur at one phase of the cycle rather than another.

3. The Basic Data

In the iron and steel, petroleum-refining, and bituminous coal industries, is there any evidence that the rate of innovation varied appreciably over the business cycle? As explained in the previous chapter, trade associations, engineering associations, and trade journals in these industries were asked to list the important processes and products first introduced between 1919 and 1958. They were also asked to rank the innovations, using total savings from new processes and total sales volume of new products as an index of importance. Usable replies were obtained from two associations and three journals, a total of about 175 innovations being listed by these respondents.[7] To determine when each of these innovations was first introduced commercially in this country, articles in trade and engineering journals were consulted and letters were sent to firms that produced and used the innovation. Members of the Carnegie Institute of Technology engineering faculty, the Bureau of Mines, and other such organizations were asked to check over the results. Ultimately, the desired information was obtained for over 80 percent of the innovations.[8] The results are contained in Tables 6.2, 6.3, and 6.4.

[6] Conference on Price Research, *Cost Behavior and Public Policy* (National Bureau of Economic Research, 1943), p. 166. This conclusion is based on fragmentary data regarding R and D expenditures, patent applications, and a case study of innovation in a farm implements firm.

[7] The distinction between a process and a product innovation may sometimes be blurred because a new technique that reduces the cost of some product may also alter it somewhat. In such cases, we asked the respondent to make a judgment as to whether the alteration was great enough for it to be considered a new product. In the case of petroleum refining, the product innovations are petrochemicals. We used a classification of important petrochemicals developed by a major oil company for its internal use, and sales in 1958 were used to rank them. Some of the classes are very broad, but there is no obvious bias. In steel, innovations in iron ore preparation, handling, etc., are excluded. In the coal industry, the innovations are all new techniques for the preparation of coal. This was the only type of innovation for which data could be obtained at all readily, and it may not be representative of all process innovations in bituminous coal. Most of the results in Tables 5.3 and 6.4 were computed from data and information appearing in *Coal Age*, a McGraw-Hill publication.

[8] We could obtain data for about 90 percent of the innovations in petroleum refining, about 50 percent of the innovations in the steel industry, and practically all of the innovations in the coal industry. The coverage is very good except for product innovations in steel, where it is quite sketchy. In the case of the product innovations in petroleum refining (i.e., petrochemicals), the innovator is defined to be the first petroleum company that produced the product commercially from a petroleum base. In a considerable number of these cases, the product had been previously produced by chemical companies. However, the petroleum company generally used a different process than its predecessors. In such cases, the innovations might have been considered as process innovations, but this would have made little difference to the final results. Note that innovation here means first commercial use by a member of *this* industry in *this* country. Of course, there may sometimes be a question about whether a particular installation was commercial or experimental. With regard to the rankings by importance, no sales data could be obtained to rank the product innovations

Table 6.3. *Dates of First Commercial Introduction of Process and Product Innovations, Petroleum-refining Industry, 1919–1958*

PROCESSES		PRODUCTS	
INNOVATION	DATE	INNOVATION	DATE
Burton-Clark cracking	1921	Isopropanol	1920
Dubbs cracking	1922	Propylene	1920
Tetraethyl lead as additive	1923	Butanol	1924
Clay treatment of gasoline	1924	Butylene	1924
Pipe stills and multi-draw bubble towers	1926	Odorants	1929
Octane number scale[1]	1926	Ammonia	1930
Solvent dewaxing of lubes	1927	Methanol	1930
Solvent extraction of lubes	1927	Aldehydes	1931
Hydrogenation	1930	Ethylene	1932
Delayed coking	1930	Naphthenic acids	1932
Propane deasphalting of lubes	1933	Ketones	1933
Catalytic polymerization	1934	Detergents	1936
Thermal polymerization	1934	Cresylic acids	1937
Desalting of crude	1935	Ethyl chloride	1938
Catalytic cracking (fixed bed)	1937	Aromatics	1940
Alkylation (H_2SO_4)	1938	Butadiene	1941
Catalytic reforming	1940	Rubber	1941
Catalytic cracking (fluid bed)	1941	Carbon black	1943
Solvent extraction of aromatics	1941	Ethanol	1943
Butane isomerization	1941	Cyclohexane	1945
Propane decarbonizing	1942	Tetramer	1945
Alkylation (HFL)	1942	Ethylene dichloride	1945
Catalytic cracking (moving bed)	1943	Trimer	1946
Pentane and hexane isomerization	1944	Diallyl phthalate polymers	1947
Platforming	1949	Glycerine	1948
Hydrogen treating	1950	Oxo alcohols	1948
Udex process	1952	Heptene	1948
Unifining	1954	Epoxy resins	1950
Fluid coking	1954	Paraxylene	1950
Molecular sieve separation	1958	Sulfur	1952
		Cumene	1953
		Styrene	1955
		Dibasic acids	1956
		Polyethylene	1956
		Resinous high-styrene copolymer	1957
		Polystyrene	1958

SOURCE: See section 3.

[1] Of course, this innovation is quite different from the others in that it is not a new type of equipment or process but a technique of measurement.

Table 6.4. *Dates of First Commercial Introduction of Process and Product Innovations, Bituminous Coal Preparation, 1919–1958*

INNOVATION	DATE
Deister tables	1919
Ruggles-Cole kiln dryer	1919
Carpenter dryer	1920
Menzies hydroseparators	1925
Rheolaveur	1925
Chance cleaner	1925
Simon-Carves washers	1928
Froth flotation	1930
Stump air-flow cleaners	1932
Link-Belt "Roto Louvre" dryer	1936
Vissac (McNally) dryer	1937
Menzies cone separators	1938
Raymond flash dryers	1939
CMI drying units	1940
Link-Belt multilouvre dryer	1946
Cyclones as thickeners	1946
Link-Belt separators	1947
Bird centrifugal filters	1947
Baughman verti-van dryer	1948
Vissac Pulso updraft dryer	1950
Eimco filter	1951
Dorrco Fluosolids machine	1955
Feldspar type jig	1956
Bird-Humboldt centrifugal dryer	1956
Wemco Fagergren flotation units	1956
Parry entrainment dryer	1958
Heyl and Patterson fluid bed dryer	1958
Continuous horizontal filter	1958

SOURCE: See section 3.

The roughness of these data should be noted at the outset. First, they suffer from the lack of a more precise way to define and evaluate an innovation. Some of these innovations could perhaps be regarded as a set of separate innovations which were introduced simultaneously, not a single innovation.

in steel. Tariff Commission data for 1958 were used for this purpose in the case of the product innovations in petroleum. In the other cases, ranks were obtained both from respondents and Carnegie (or Bureau of Mines) personnel; and the average of these ranks was used. These data are obviously very rough, but it is noteworthy that in each case the independent rankings were highly correlated, indicating a considerable amount of agreement. The weight we used is the innovation's average rank divided by a constant that makes the average value of the weights equal to one.

If they were, the results using unweighted data would depend on how many elements were recognized as separate innovations in each case. Although the weighted data used below should help to eliminate this problem, the weights obviously are not very reliable. Second, for some purposes, the results may suffer from the fact that only innovations which exceeded a minimum level of economic importance were included. The minor improvements, which can accumulate over a period of years to become very important, are excluded entirely.[9] Third, biases may be present because early innovations tend to be forgotten or because the importance of recent innovations is particularly difficult to judge. Although the pooling of many individual judgments should have eliminated many such errors, some undoubtedly persist.[10]

Despite these limitations, it seems worthwhile to use these data for exploratory purposes. There is no reason to believe that the errors in the data are correlated with the business cycle or with the variables contained in the investment function in section 6. Thus, although there is likely to be considerable "noise," there is no reason to expect much bias in the results. Moreover, if we are interested in studying the timing of innovation, there is little choice, since these data seem to be the best, if not the only, description currently available of the distribution of significant innovations over time.

4. The Empirical Results

In each industry, we assume that

$$n_i = \beta_0 + \beta_1 u_i + \beta_2 u_i^2 + z_i, \tag{1}$$

where n_i is the number of innovations occurring in year i, u_i is the average percent of capacity utilized by the industry during year i, and z_i is a random error term. We use the percent of an industry's capacity that is utilized as a measure of where a particular year lies in the business cycle. This is reasonable, but rough. No trend term is included in equation (1) because in every case when we added it it turned out to be statistically nonsignificant, and the estimates of the β's remained close to the values given below.[11]

[9] Of course, the results are also of little use in testing hypotheses, like Schumpeter's, which are concerned primarily with innovations that resulted in the formation of new industries and the revolutionizing of old ones. Very few of the innovations in Tables 6.2, 6.3, and 6.4 are that important.

[10] The results are limited in scope. Some attempt was made to extend the coverage to other industries but the necessary data could not be obtained. Moreover, some bias may result from our failure, noted above, to determine when some of the innovations listed by the respondents were first introduced. However, this could be of importance only for product innovations in steel. Also, the weights seem to underestimate the importance of a few of the newer steel techniques like continuous casting.

[11] Since the standard errors of the trend terms are quite large, it is impossible to say much about the size of these terms. Because the estimates are statistically nonsignificant, it does not follow, of course, that they are actually zero.

Some function of lagged R and D expenditures by the industry in question and its suppliers might also be used as an independent variable in equation (1). Lack of data prevented me from trying it.

When we used data regarding u_i from the American Iron and Steel Institute, the American Petroleum Institute, and the Bureau of Mines, we obtained least-squares estimates of the β's.[12] Although these estimates are unbiased, they may be somewhat inefficient, because z_i may be heteroscedastic.[13] These results are presented in Table 6.5.

In the case of process innovations, $\hat{\beta}_1$ is always positive and $\hat{\beta}_2$ is always negative; and both are generally statistically significant. Thus, up to some point, increases in the rate of utilization of capacity seem to be associated with increases in the rate of occurrence of process innovations; beyond that point, they are associated with decreases in their rate of occurrence. Apparently, the maximum rate of innovation can be expected when these industries are operating at about 67 percent (steel), 73 percent (petroleum), and 75 percent (coal) of capacity. If they are operating at 50 or 100 percent of capacity, the rate of innovation is only about one-third of the maximum rate.[14]

Why is this the case? According to executives in these industries, new processes are unlikely to be introduced when an industry is operating at low levels of capacity utilization, because the risks involved seem particularly great under those circumstances, profits being slim and the future seeming particularly uncertain. On the other hand, when an industry is operating at very high levels of capacity utilization, there is some reluctance to innovate because it will interfere with production schedules: there is little unutilized capacity that can easily and cheaply be used for "experimental" purposes.[15]

As for product innovations, the results in Table 6.5 provide no evidence that the rate of innovation varies appreciably over the business cycle. None of the estimates of β_1 or β_2 is statistically significant. Of course, this may be due to "noise" in the basic data, which are probably not as accurate as those for process innovations.

[12] In iron and steel, the figure used is ingot production as a percent of ingot capacity. In petroleum, the figure is annual average (daily) crude runs to stills as a percent of the (average of January 1 and December 31) refining capacity figure given by the Bureau of Mines. In coal, the figure comes from the Bureau of Mines estimates.

[13] One might expect that the variance of n_i is directly related to its mean. If so, the standard errors in Table 6.5 will be somewhat in error. Also, it would have been better to have used data for shorter periods than a year but no such data could be obtained.

[14] According to the estimates in Table 6.5 the expected number of innovations at 50 percent of capacity is 84 percent (steel process), 0 percent (petroleum process), and 21 percent (coal preparation) of the maximum expected number. At 100 percent of capacity, the expected number of innovations is 38 percent (steel process), 0 percent (petroleum process), and 21 percent (coal preparation) of the maximum expected number. Needless to say, all of these estimates are subject to very large sampling errors. Note that if the war years (which account for many of the cases where these industries were operating at very high levels of capacity utilization) are omitted, the same general sorts of relationships between n_i and u_i seem to persist.

[15] If this argument is true, it is surprising that the maximum rate of innovation occurs at a fairly low level of utilization of capacity. (On this basis, I would suppose that it would occur at perhaps 80 to 90 percent of capacity.) However, the reason for this may lie in deficiencies in the capacity measures and in sampling errors.

Table 6.5. *Estimates of β's and φ's, Iron and Steel, Petroleum-refining, and Bituminous Coal Industries, Process and Product Innovations, 1919–1958*

INDUSTRY AND TYPE OF DATA	PARAMETERS[a]						
	β_0	β_1	β_2	ϕ_0	ϕ_1	ϕ_2	ϕ_3
			Process Innovations				
Iron and steel							
Weighted[b]	−1.59	0.08[d]	−0.0006[d]	−1.56	0.08[d]	−0.0006[d]	−0.06
		(.05)	(.0004)		(.05)	(.0004)	(.39)
Unweighted	−1.36	.07	−.0005	−1.35	.07	−.0005	−.06
		(.05)	(.0004)		(.05)	(.0004)	(.36)
Petroleum refining							
Weighted[b]	−16.21	.47[e]	−.0032[e]	−16.30	.47[e]	−.0032[e]	.04
		(.22)	(.0014)		(.22)	(.0014)	(.27)
Unweighted	−11.09	.34[d]	−.0024[d]	−10.94	.33	−.0023[d]	−.13
		(.21)	(.0014)		(.21)	(.0014)	(.26)
Bituminous coal							
Weighted[b]	−8.89	.26[e]	−.0018[e]	−9.48	.28[e]	−.0019[e]	−.14
		(.13)	(.0009)		(.14)	(.0009)	(.35)
Unweighted	−9.93	.29[e]	−.0020[e]	−10.55	.31[e]	−.0021[e]	−.15
		(.12)	(.0008)		(.13)	(.0009)	(.31)
			Product Innovations				
Iron and steel[c]							
Unweighted	.17	.02	−.0002	.01	.02	−.0002	.32
		(.03)	(.0002)		(.03)	(.0002)	(.22)
Petroleum refining							
Weighted[b]	2.32	−.06	.0005	2.35	−.06	.0005	−.02
		(.28)	(.0019)		(.29)	(.0019)	(.35)
Unweighted	10.77	−.29	.0020	10.66	−.29	.0020	.09
		(.24)	(.0016)		(.24)	(.0016)	(.29)

SOURCE: See sections 3 and 4.

[a] See equation (1) for the definition of the β's, and equation (2) for the definition of the φ's.

[b] The weights are discussed in note 8.

[c] No weights could be obtained for product innovations in steel.

[d] Statistically significant at 0.10 level.

[e] Statistically significant at 0.05 level.

Thus far, no attempt has been made to distinguish years in which an industry's rate of capacity utilization is falling from those in which it is rising. To include this factor, we assume that

$$n_i = \phi_0 + \phi_1 u_i + \phi_2 u_i^2 + \phi_3 C_i + z'_i, \tag{2}$$

where C_i equals one if $u_i > u_{i-1}$ and zero otherwise. Of course, this is but one of many ways in which this factor might have been introduced into the model. Estimates of the ϕ's are shown in Table 6.5. From these results, one can find no persistent tendency for the rate of innovation to be higher or lower, depending on whether or not the rate of utilization of the industry's capacity has increased or decreased from the previous year. In some cases ϕ_3 is positive; in some cases it is negative; and in every case it is statistically nonsignificant.[16]

To summarize, the rate of process innovation in these industries varied substantially over the business cycle, and it reached a maximum when about 75 percent of the industry's capacity was utilized. The rate of product innovation did not vary significantly over the business cycle, but the tests are not very powerful. Given the level of capacity utilization, there is no evidence that it made a difference whether or not it was falling or rising.

5. The Investment Function

This section attempts to incorporate the timing of innovation into a simple econometric model of investment behavior. Suppose that firms decide at the beginning of each year how much productive capacity they will build during the year and that this new capacity is completed and operable at the beginning of the next year. (The assumption that it takes one year to construct new facilities is made because, according to Modigliani [111], this is generally the case in iron and steel and petroleum refining, the two industries used in the empirical work. Other gestation periods besides one year could be used instead, if they seemed more appropriate.) If the firm decides to build no new capacity during year t, suppose that its capacity at the beginning of year $(t + 1)$ would be $(1 - y)k_{t-1}$, where k_{t-1} is its capacity at the beginning of year t and y is the proportion of its facilities that "wear out" during the year.

In deciding at the beginning of year t how much capacity to build, suppose that the firm forecasts its output during year $(t + 1)$, this forecast equaling ηo_{t-1}, where o_{t-1} is its actual output in year $(t - 1)$. Suppose that there is associated with this forecasted output for year $(t + 1)$ a certain desired amount of capacity at the beginning of year $(t + 1)$, this desired capacity level being $\theta\eta o_{t-1}$. And suppose that the firm builds enough capacity during the year to eliminate 100 π percent of the difference between the desired capacity level at the beginning of year $(t + 1)$ and the amount of capacity its existing plant (at the beginning of year t) would represent at the beginning of year $(t + 1)$.

If these assumptions hold and if $\theta\eta o_{t-1} > (1 - y)k_{t-1}$ for all firms, it follows that

$$D_t = \pi[\theta\eta 0_{t-1} - (1 - y)K_{t-1}], \tag{3}$$

[16] The estimates of ϕ_3 tend to be negative for processes and positive for products, but this may well be due to chance.

where D_t is the total amount of capacity built by the industry during year t, 0_{t-1} is the total industry output during year $(t - 1)$, and K_{t-1} is the total capacity of the industry at the beginning of year t. Moreover, if P_t is the cost during year t of building a unit of capacity, the industry's total expenditures during year t to expand capacity and replace "worn-out" equipment are:

$$(4) \qquad E_t = P_t\pi[\theta\eta 0_{t-1} - (1 - y)K_{t-1}].$$

In addition, firms invest in new processes and in facilities required to produce new products. To the extent that these innovations are incorporated in new capacity, such expenditures are already included in the cost of constructing capacity in equation (4). But to the extent that they are used to up-grade and re-equip old capacity to make it more efficient and to extend the spectrum of products derivable from it, the expenditures are not included. To complete the model, assume that it takes x years for all the industry's old capacity to be equipped with an innovation. During each year of this period (beginning from the year when it was first commercially introduced), assume that $100/x$ percent of the capacity existing at the beginning of the year is so equipped, the cost per unit of capacity being w_j dollars for the jth innovation.[17] Thus,

$$(5) \qquad E'_t = \overline{w}_t x^{-1} K_{t-1} \sum_{s=t-x-1}^{s=t-1} n(s),$$

where E'_t is the expenditure during year t on the introduction of innovations into old capacity, $\overline{w}_t$ is the average value of w_j for innovations first occurring

[17] Of course, the assumption that these expenditures go on at a constant rate (in the sense described above) is a rough simplification that is adopted as a matter of convenience. A more sophisticated model of the diffusion process is presented in Chaps. 7 to 9. The simpler, but cruder, assumption seems good enough for present purposes. Note too that all of the existing capacity need not be equipped with the new technique; for example, it makes no difference if every other unit of capacity is so equipped eventually. All that we need to do is to interpret $\overline{w}_t$ as the average cost of installing these innovations, where the average is taken over all units of capacity, whether they use the innovation or not. So long as $\overline{w}_t$ is reasonably stable over time, this makes no difference in the following discussion. Note that since K_{t-1} will vary over time, the expenditures on a given innovation will vary over time too. Furthermore, because K_{t-1} does not remain constant, the total amount of capacity equipped with the innovation at the end of the x year period will not equal the total amount of capacity existing at the beginning of the x year period. This problem does not seem too important, but it should be noted. In fact, such a difference could occur because some new capacity built during the period was not equipped with the innovation. One way of eliminating this problem might be to use $\sum_{s=t-x-1}^{t-1} n(s)K_{s-1}$ in equation (5) rather than $K_{t-1} \sum_{s=t-x-1}^{t-1} n(s)$. The results in equations (8) and (9) turn out to be essentially the same in petroleum but different in steel if this variable is used. The estimates of a_3, a_1, and a_2 are 0.46 (0.24), 10.6 (5.8), and −11.3 (6.1) in steel and 1.46 (0.99), 388 (130), and −288 (112) in petroleum, the numbers in parentheses being the standard errors. The trouble with this procedure is that it assumes that an innovation is accepted immediately for all new capacity —which is far from the case.

from year $(t - x - 1)$ to year $(t - 1)$, and $n(s)$ is the number of innovations first occurring in year s.

If all expenditures on plant and equipment are included in E_t or E'_t, it follows that

$$(6) \qquad I_t = P_t\pi[\theta\eta 0_{t-1} - (1 - y)K_{t-1}] + \overline{w}_t x^{-1} K_{t-1} \sum_{s=t-x-1}^{s=t-1} n(s),$$

where I_t is the total expenditure by the industry on plant and equipment during year t.[18]

6. Tests of the Model

This section uses 1946–1959 data regarding the iron and steel and petroleum-refining industries to test this model and to estimate some of the parameters. In carrying out these tests, it is convenient to make two further assumptions. First, assume that P_t was constant in real terms during 1946–1959 in these industries. Because the time period is relatively short and the innovations occurring then were unlikely to have a great effect on the capital coefficient, this assumption is probably not too unsatisfactory.[19] Second, assume that $\overline{w}_t$ was uncorrelated during this period with 0_{t-1}, K_{t-1}, and $\sum_{s=t-x-1}^{t-1} n(s)$. On a priori grounds, it is hard to see why such correlations should exist. If these assumptions hold, it follows from equation (6) that

$$(7) \qquad I_t = a_1 0_{t-1} + a_2 K_{t-1} + a_3 K_{t-1} \sum_{s=t-x-1}^{s=t-1} n(s) + g_t,$$

where $a_1 = P\pi\theta\eta$, $a_2 = -P\eta(1 - y)$, $a_3 = \overline{\overline{w}}x^{-1}$, $\overline{\overline{w}}$ is the mean of $\overline{w}_t$, and g_t is a random error term.[20]

To estimate the a's, data are needed regarding I_t, 0_{t-1}, K_{t-1}, x, and $\sum_{s=t-x-1}^{t-1} n(s)$. In petroleum, estimates of I_t were obtained by deflating the

[18] Of course, the model is oversimplified in many respects. Some of the oversimplifications are discussed in more detail in section 6.

[19] D. Creamer's data ("Postwar Trends in the Relation of Capital to Output in Manufactures," *American Economic Review* [May 1958]) do not indicate any substantial change during this period in the capital-output ratio in these industries. J. Hodges's study of petroleum refining also seems to bear out this assumption ("A Report on the Calculation of Capital Coefficients for the Petroleum Industry," *Problems of Capital Formation* [Princeton, 1957]).

[20] Since $g_t = (\overline{w}_t - \overline{\overline{w}})K_t \sum_{s=t-x-1}^{t-1} n(s)$, it is not homoscedastic. But if all terms in equation (7) are divided by $K_t \sum_{s=t-x-1}^{t-1} n(s)$, the resulting residual is homoscedastic, and if estimates of the a's are made using these new variables, the results are best linear unbiased estimates.

Chase-Manhattan Bank's estimates of the industry's gross investment in refineries and chemical plants by the *Engineering News Record's* index of construction costs. In steel, estimates of I_t were obtained by deflating OBE-SEC data on expenditures on plant and equipment by the same index. Estimates of 0_{t-1} and K_{t-1} (expressed in millions of barrels of crude oil or millions of tons of ingots) were obtained from the American Iron and Steel Institute and the American Petroleum Institute. Alternative values of x were tried, and the value was used that provided the best fit—23 years in iron and steel and 20 years in petroleum refining. These estimates of x seem quite reasonable, when compared with more direct measures of the rate of diffusion

Table 6.6. *Values of I_t, 0_{t-1}, K_{t-1}, $K_{t-1} \sum_{s=t-x-1}^{t-1} n(s)$, Iron and Steel and Petroleum-refining Industries, 1946–1959*

YEAR	IRON AND STEEL				PETROLEUM			
	I_t	0_{t-1}	K_{t-1}	$K_{t-1} \sum_{t-24}^{t-1} n(s)$	I_t	0_{t-1}	K_{t-1}	$K_{t-1} \sum_{t-21}^{t-1} n(s)$
1946	907.4	79.7	91.9	3032.7	453.7	4.7	5.3	201.4
1947	974.0	66.6	91.2	3100.8	610.7	4.7	5.6	218.4
1948	1051.8	84.9	94.2	3202.8	817.4	5.1	6.0	228.0
1949	784.2	88.6	96.1	3171.3	552.6	5.6	6.4	249.6
1950	740.4	78.0	99.4	3479.0	339.9	5.3	6.7	268.0
1951	1391.4	96.8	104.2	3542.8	377.5	5.7	7.0	294.0
1952	1667.7	105.1	108.6	3692.4	518.8	6.5	7.3	277.4
1953	1267.0	93.2	117.5	3877.5	706.8	6.7	7.6	296.4
1954	754.0	111.6	124.3	3977.6	800.0	7.0	8.0	304.0
1955	822.7	88.3	125.8	4151.4	796.0	7.0	8.4	319.2
1956	1151.6	117.0	128.4	4237.2	749.3	7.5	8.6	318.2
1957	1434.7	115.2	133.5	4539.0	826.1	7.9	9.1	345.8
1958	988.3	112.7	140.7	4643.1	601.2	7.9	9.4	357.2
1959	815.7	85.3	147.6	4870.8	413.4	7.6	9.8	372.4

SOURCE: See section 6.

SYMBOLS: I_t is the deflated expenditure on plant and equipment during year t (in millions of 1954 dollars), 0_{t-1} is physical output in year $(t - 1)$, K_{t-1} is capacity at the end of year $(t - 1)$, and $n(s)$ is the number of significant innovations occurring in year s. For the units in which 0_{t-1} and K_{t-1} are measured, see section 6.

in these and other industries.[21] Finally, Tables 6.2, 6.3, and 6.4 were used to determine $K_{t-1} \sum_{s=t-x-1}^{t-1} n(s)$.

The results are shown in Table 6.6. From these data, we obtained

[21] See H. Jerome, *Mechanization in Industry* (National Bureau of Economic Research, 1934), p. 313; and Chap. 7 of this book.

least-squares estimates of the a's, after dividing both sides of equation (7) by $K_{t-1} \sum_{s=t-x-1}^{t-1} n(s)$ to insure homoscedasticity of the residuals. Division by $K_{t-1} \sum_{s=t-x-1}^{t-1} n(s)$ also reduces substantially the problem of multicollinearity. In the iron and steel industry, the results are

$$(8)\quad \frac{I_t}{K_{t-1} \sum_{s=t-24}^{t-1} n(s)} = \underset{(0.75)}{1.80} + \underset{(5.3)}{14.5} \frac{0_{t-1}}{K_{t-1} \sum_{s=t-24}^{t-1} n(s)} - \underset{(25.3)}{62.7} \frac{1}{\sum_{s=t-24}^{t-1} n(s)} \cdot (r = 0.67)$$

In the petroleum industry, the results are

$$(9)\quad \frac{I_t}{K_{t-1} \sum_{s=t-21}^{t-1} n(s)} = \underset{(0.72)}{1.38} + \underset{(124)}{387} \frac{0_{t-1}}{K_{t-1} \sum_{s=t-21}^{t-1} n(s)} - \underset{(109)}{305} \frac{1}{\sum_{s=t-21}^{t-1} n(s)} \cdot (r = 0.62)$$

In both cases, the error term is omitted and the numbers in parentheses under the regression coefficients are their standard errors.

Although the fits are not particularly good (for reasons discussed in section 7), the results are encouraging in all other respects. All of the regression coefficients have the expected signs, all are statistically significant at the 0.05 level; and there is no evidence of autocorrelation in the residuals, the Durbin-Watson statistic being 1.66 in iron and steel and 1.67 in petroleum. More important, the estimates of a_1 and a_2 are not significantly different from what would be expected on the basis of independent estimates of P, π, θ, η, and y. Indeed, the differences generally are quite small.[22]

[22] To obtain independent estimates of θ, we used the desired rates of utilization of capacity provided by McGraw-Hill, *Business Plans for Expenditures on Plant and Equipment* (annual), adjusted to take account of the fact that the McGraw-Hill rates of utilization are different from ours. In particular, we divided the desired rate as reported by McGraw-Hill by the average ratio of McGraw-Hill's actual rate to our actual rate. The resulting estimates of θ are 1.23 in steel and 1.18 in oil. H. Chenery (in "Overcapacity and the Acceleration Principle," *Econometrica*, Jan. 1952) provides estimates of π—0.07 in steel and 0.65 in oil. According to testimony in the Bethlehem-Youngstown case, a reasonable value of P in steel is about $200. According to J. McLean and R. Haigh, *The Growth of Integrated Oil Companies* (Cambridge: Harvard University Press, 1954), p. 560, a reasonable value of P in oil is about $545. Assuming that η is 1.03 in steel and 1.05 in oil and that y is 0.05 in both industries, we obtain $18 (steel) and $435 (oil) as independent estimates of a_1, and −$14 (steel) and −$336 (oil) as independent estimates of a_2. None of the estimates in equations (8) and (9) differs significantly from these figures, and except for a_2 in steel the differences are fairly small.

Note that it is appropriate in equations (8) and (9) to use one-tailed tests, since the signs

Since results like those in equations (8) and (9) were computed for a fairly wide range of values of x (15 to 23 in petroleum and 15 to 25 in steel), it is possible to determine how sensitive the regression coefficients are to changes in x. Regardless of which value of x I used, all of the regression coefficients maintained the expected signs, there being only one exception in sixty cases. The estimates of a_3 in iron and steel and a_1 and a_2 in petroleum refining seemed to be quite insensitive to almost all changes in x, but the other three coefficients varied considerably. Regardless of the value of x, most of the coefficients were statistically significant.[23]

7. Implications and Discussion

These findings have at least three implications for studies of investment behavior. First, they suggest that, using even a crude model which includes the timing of innovation, one can do a significantly better job of explaining variation in investment than if one uses the flexible "capacity accelerator" alone. If the flexible "capacity accelerator" employed by Chenery [24], Goodwin [42], Modigliani [111], and others were used here, only 0_{t-1} and K_{t-1} would be included as exogenous variables. Thus, since $K_{t-1} \sum_{s=t-x-1}^{t-1} n(s)$ has a statistically significant effect on I_t (0_{t-1} and K_{t-1} held constant), it follows that a significantly better explanation of the behavior of I_t can be achieved in these cases by including the timing of innovation as well.[24]

Second, the results enable us to estimate very roughly the average annual expenditures made in these industries to fit out existing capacity with one of these innovations. Judging by equations (8) and (9), one can be quite sure (0.975 fiducial probability in steel and 0.90 fiducial probability in petroleum) that the average expenditure of this sort on one of the (relevant) innovations

---

of the coefficients are given by the model. Of course, in judging the significance of the results in these equations, one should take account of the fact that x was fitted experimentally and that the estimated standard errors are probably too low for this reason.

[23] We obtained results for $x = 15 - 23$ in each industry, since according to H. Jerome (*Mechanization in Industry* [National Bureau of Economic Research, 1934]) and Chap. 7, this seemed to be the relevant range. Because the maximum correlation coefficient was obtained when $x = 23$ in the steel industry, we also obtained results for $x = 24$, 25 in the industry to see if the fit could be improved further. Beginning with $x = 15$ and continuing in order, the estimates in steel of a_1 were 12.2, 8.7, 8.6, 12.6, 13.8, 11.5, 11.1, 11.2, 14.5, 13.6, 12.8; the estimates of a_2 were −8.2, −10.4, −8.9, −8.4, −8.3, −9.9, −9.0, −36.6, −62.7, −47.0, and −29.2; the estimates of a_3 were 0.35, 0.55, 0.46, 0.29, 0.23, 0.34, 0.30, 1.14, 1.80, 1.31, and 0.79. In petroleum, the estimates of a_1 were 346, 364, 550, 400, 400, 387, 375, 392, and 382; the estimates of a_2 were −268, −280, −924, −235, −265, −305, −280, −290, −272; the estimates of a_3 were 1.91, 1.68, 16.0, −0.60, 0.20, 1.38, 0.97, 0.81, and 0.60. Practically all of the estimates of a_1 and a_2, and about half of the estimates of a_3, are significant at the 0.10 level or better.

[24] There are considerable problems in using equation (7) to forecast the level of investment in an industry. Here we take as given the number of innovations occurring in each year during 1919–1958 that exceeded a certain level of importance. In forecasting, we would have to identify the number occurring in the immediately previous years—often a very difficult task.

by the entire industry in 1946 was *at least* \$13.8 million in steel and \$2.1 million in petroleum. In 1957, one can be equally sure that it was *at least* \$20.0 million in steel and \$3.6 million in petroleum. All expenditures are expressed in 1954 dollars.[25]

If we go a step further, the total expenditure of this sort by the steel industry was *at least* \$455 million in 1946 and \$681 million in 1957. In petroleum, it was *at least* \$81 million in 1946 and \$138 million in 1957. In both industries, these figures are probably very conservative. They are lower than McGraw-Hill's estimates [75] of the expenditures devoted to replacement and modernization rather than expansion. Of course, this is what one would expect, since the McGraw-Hill figures are estimates whereas ours are lower-bounds, and theirs include replacement whereas ours do not.[26]

Third, the findings suggest that innovation-induced investment to refurbish existing capacity would be of a cyclical nature, even if the industry operated at a constant level of capacity utilization, if K_{t-1} were constant, and if the expected value of $n(s)$ were constant over time. This is because such investment would be, under the hypothesized circumstances (approximately) proportional to $\frac{1}{x}\sum_{s=t-x-1}^{t-1} n(s)$, a moving average of identically distributed random variables. According to the Yule-Slutsky theorem [153], such a moving average will exhibit a cyclical behavior.[27]

Finally, it may be worthwhile to point out some of the reasons why equations (8) and (9) provide only a moderately good fit. First, the assumption that firms form sales expectations on the basis of a simple extrapolation of the previous year's output is obviously quite weak. Although this assumption is often used, and sometimes with considerable success [67], it is probably only a poor approximation in most industries. Second, the measures of capacity and output are incomplete, since they pertain only to ingot production and refinery throughput. Although these are key steps in the production processes, other steps like rolling (in steel) and petrochemicals (in petroleum) are also important. Third, other factors like the profitability and liquidity of the firms, the cost of capital, and the age of existing equipment are likely to influence I_t. Their inclusion undoubtedly would improve the results.[28]

[25] Assume that the error term in equation (8) was normally distributed; then the probability is 0.975 that a_3 is greater than $\hat{a}_3 - 2.20\,\hat{\sigma}_{a_3}$, where $\hat{\sigma}_{a_3}$ is the standard error of a_3. That is, an interval from ∞ to $\hat{a}_3 - 2.20\,\hat{\sigma}_{a_3}$ will include a_3 in 97.5 percent of the cases. Inserting the values of $\hat{a}_3$ and $\hat{\sigma}_{a_3}$ in equation (8) into this expression and multiplying by K_{t-1}, we obtain the figures in the text for the steel industry. The procedure for the petroleum industry is the same, except that 1.36 is used rather than 2.20.

[26] These figures were obtained by multiplying the results of the previous footnote by $\sum_{s=t-x-1}^{t-1} n(s)$. Another reason why McGraw-Hill's estimates should be higher is that we do not include all innovations on our lists.

[27] Of course, Frisch pointed out many years ago that a mechanism of this sort might be at work.

[28] It may also be advisable to let π be a function of the size of the gap between the actual and desired capital stock. See Chap. 2.

8. Successful Innovation and the Growth of Firms

How much of an impact does a successful innovation have on a firm's growth rate?[29] In Chapter 5, we listed the firms that were first to introduce

Table 6.7. *Average Annual Growth Rates of Successful Innovators and Other Firms (of Comparable Initial Size), Computed Values of $\bar{e}$ and $\bar{d}$, and Regressions (Excluding Innovators) of ln $S_{ij}^{t+\Delta}$ on ln S_{ij}^{t}, Iron and Steel and Petroleum-refining Industries, Selected Periods*

	STEEL				PETROLEUM			
ITEM	1916–1926	1926–1935	1935–1945	1945–1954	1921–1927	1927–1937	1937–1947	1947–1957
Average annual growth rate (percent):								
Innovators	13.7	6.5	3.4	3.2	13.1	7.9	3.6	6.7
Other firms	3.7	3.3	2.0	2.4	6.6	4.1	3.6	4.2
Computed value of[a]:								
$\bar{e}$ (percent)	—	0.7	0.7	—	—	4.2	−2.5	−2.8
$\bar{d}$ (percent)	—	3.9	5.2	—	—	5.7	3.6	13.4
Regression (excluding innovators) of ln $S_{ij}^{t+\Delta}$ on ln S_{ij}^{t}:								
Intercept (a_i)	1.68	.55	1.34	.18	1.10	1.68	.41	1.27
Slope (b_i)	.88	.97	.90	1.01	.93	.84	.98	.90

SOURCE: See Chap. 5 and E. Mansfield, "Entry, Gibrat's Law, Innovation, and the Growth of Firms," *American Economic Review*, Dec. 1962.

SYMBOLS: S_{ij}^{t} is the size of the jth firm in the ith industry at time t, $S_{ij}^{t+\Delta}$ is its size at time $t + \Delta$; $\bar{e}$ is the average value of e_j, where e_j is the difference between the average annual growth rate of the jth innovator during the period from time t to time t_j and that of "other firms" of equivalent size (at time t) during the same period; and $\bar{d}$ is the average value of d_j, where $e_j + d_j$ is the difference between the average annual growth rate of the jth innovator during the period from time t_j to time $t + \Delta$ and that of "other firms" of equivalent size (at time t) during the same period. See note 32 for the way in which the regressions described here are used to estimate the figures in the second row of this table.

[a] No figures are computed in cases in which there were only a few innovators. See note 32 for a discussion of the derivation of these figures.

the significant new processes and products in steel and petroleum. A comparison of their growth rates with those of other comparable firms and with their own preinnovation growth rates should help to indicate how great the

[29] This section draws on part of my "Entry, Gibrat's Law, Innovation, and the Growth of Firms," *American Economic Review* (Dec. 1962).

payoff is (in terms of growth) for a successful innovation. For each period for which we have data, Table 6.7 estimates the average annual growth rate of (1) firms that carried out significant innovations during the period, and (2) other firms that were equal in size to the successful innovators at the beginning of the period. There is a marked difference between the two groups. In every time interval and in both industries, the successful innovators grew more rapidly than the others; and in some cases, their average rate of growth was more than twice that of the others.[30]

When each innovator is considered separately, the difference between its growth rate and the average growth rate of other comparable firms seems to have been inversely related to its size. As one would expect, a successful innovation had a much greater impact on the growth rate of a small firm than on that of a large firm. The fact that fewer of the successful innovators in more recent periods were small firms probably accounts in part for the decrease over time in the average difference (in Table 6.7) between the two groups.[31]

To what extent did the innovator grow more rapidly because of certain factors associated with the innovation, not because of the innovation itself? Each growth rate in Table 6.7 pertains to the entire period indicated at the top of each column, whereas the innovations occurred at some time within the period. Consider the period from time t to time $t + \Delta$. Suppose that the jth successful innovator in this period introduced its innovation at time t_j, that its average annual growth rate from time t to time t_j exceeded that of other comparable firms by e_j, and that its average annual growth rate from time t_j to time $t + \Delta$ exceeded that of other comparable firms by $e_j + d_j$. What were the average values of e_j and d_j? If the innovators grew more rapidly than other firms because of certain characteristics associated with the innovation, but not because of the innovation itself, and if these characteristics had approximately the same effect throughout the period, the average value of d_j would be expected to be zero.

Letting $S_j^{t+\Delta}$ be the size (that is, capacity in tons or barrels) at time $t + \Delta$

[30] Note four things: First, we are not comparing innovators with noninnovators, since some of the "other firms" may have been unsuccessful innovators. Because we can only include successful innovators (the data being what they are), it is not surprising that they have higher growth rates, and we are much more interested in the size of the difference than in its existence. Second, some of the innovators introduced more than one innovation during the period. Thus, the difference in growth rates is not due entirely to a single innovation. But in the subsequent analysis (involving $\bar{d}$) only cases involving a single innovation are included. Third, it would be interesting to see how an innovation's effects depended on its character, but we have too little data to attempt this. Fourth, the way in which the average annual growth rate of the "other firms" in Table 6.7 was computed is described in note 32.

[31] If the innovators in steel are divided into two groups—those above 1,000,000 tons and those less than (or equal to) 1,000,000 tons at the beginning of the period—the average difference between their growth rates and the growth rates of other comparable firms differs considerably between the groups. Among the larger firms, the average difference is generally about 0.5 points whereas it is 3 to 10 points among the smaller ones. Similarly, if the innovators in petroleum are divided into two groups—those above 32,000 barrels and those less than (or equal to) 32,000 barrels at the beginning of the period—the average difference is practically zero among the larger firms but 6 to 24 points among the smaller ones.

of the jth innovator and $Q_j^{t+\Delta}$ be the average logarithm of the sizes at time $t + \Delta$ of the other firms that were equal in size to the jth innovator at time t, one can show that

$$\text{(10)} \qquad (ln\ S_j^{t+\Delta} - Q_j^{t+\Delta})/\Delta = e_j + [1 - (t_j - t)/\Delta]d_j.$$

To see this, consider the kth "other firm" of the same size as the jth innovator at time t. If r_{1k} is its average rate of growth between time t and time t_j, r_{2k} is its average rate of growth between time t_j and time $t + \Delta$, and $S_{jk}^{t+\Delta}$ is its size at time $t + \Delta$,

$$ln\ S_{jk}^{t+\Delta} = ln\ S_j^t + r_{1k}(t_j - t) + r_{2k}(t + \Delta - t_j).$$

Thus, if r_1 is the average value of r_{1k} and r_2 is the average value of r_{2k},

$$Q_j^{t+\Delta} = ln\ S_j^t + r_1(t_j - t) + r_2(t + \Delta - t_j).$$

But by the definitions of e_j and d_j,

$$ln\ S_j^{t+\Delta} = ln\ S_j^t + (r_1 + e_j)(t_j - t) + (r_2 + e_j + d_j)(t + \Delta - t_j).$$

Thus,

$$ln\ S_j^{t+\Delta} - Q_j^{t+\Delta} = e_j\Delta + d_j(t + \Delta - t_j),$$

and equation (10) follows.

If $\bar{e}$ and $\bar{d}$ are the average values of e_j and d_j, and if we assume that $(e_j - \bar{e})$ and $(d_j - \bar{d})$ are statistically independent of $(t_j - t)/\Delta$, it follows that

$$\text{(11)} \qquad (ln\ S_j^{t+\Delta} - Q_j^{t+\Delta})/\Delta = \bar{e} + [1 - (t_j - t)/\Delta]\bar{d} + W_j,$$

where W_j can be treated as a random error term. Using equation (11), we can apply least-squares to obtain $\bar{e}$ and $\bar{d}$.[32]

[32] How did we estimate $Q_j^{t+\Delta}$? If the innovator was smaller than 1,000,000 tons (steel) or 64,000 barrels (petroleum), we used the following technique to estimate the average annual growth rate of the other firms of its initial size. We assumed that, for the jth "other firm" in the ith industry, $ln\ S_{ij}^{t+\Delta} = a_i + b_i\ ln\ S_{ij}^t + Z_{ij}'''$, where Z_{ij}''' is a random error term. An equation of this form fits the data for the smaller firms quite well. We then obtained least-squares estimates (shown in Table 6.7) of a_i and b_i; and taking each innovator, we used this regression to estimate the average value of $ln\ S_{ij}^{t+\Delta}$ for the "other firms" corresponding to its value of S_{ij}^t. Deducting its value of $ln\ S_{ij}^t$ from this computed average value and dividing the result by Δ, we obtain an estimate of the average annual growth rate of "other firms" of the same original size as this innovator.

If the innovator was larger than 1,000,000 tons or 64,000 barrels, we had to use another method because the regressions do not always fit the larger firms very well. In these cases, we used the average annual growth rate of the "other firms" larger than 1,000,000 tons or 64,000 barrels. Finally, to obtain the figures in the second row of Table 6.7, we took the resulting average growth rate for the "other firms" corresponding to each innovator during the period (whether or not it was above 1,000,000 tons or 64,000 barrels) and averaged them. In computing $\bar{e}$ and d, innovators that introduced more than one innovation had to be excluded (except in a few cases where the innovations were all introduced at the same

The results (in Table 6.7) indicate that $\bar{d}$ was always positive, but that the sign of $\bar{e}$ varied. This means two things. First, in the period immediately before they introduced the innovations, there was no persistent tendency for the successful innovators to grow more rapidly than other comparable firms. In some cases they grew more rapidly, but in others they did not. Thus, their higher growth rate cannot be attributed to their preinnovation behavior. Second, in the period after they introduced the innovations their mean growth rate consistently exceeded that of other comparable firms by more than it had before their introduction—which is what one would expect.

Finally, if one makes the crude assumption that the preinnovation difference in average growth rate between successful innovators and other firms would have been maintained from time t to time $t + \Delta$ if the innovations had not been introduced, $\bar{d}$ measures the average effect of these successful innovations on a firm's growth rate during the relevant period. Based on this assumption, their average effect was to raise a firm's growth rate by 4 to 13 percentage points, depending on the particular time interval and industry. In view of the widespread interest in measures of the payoff from successful innovation, these estimates, despite their crudeness, should be useful.[33]

9. Summary and Conclusions

The principal conclusions of this chapter are as follows. First, the available evidence for major innovations indicates that the average lag between invention and innovation is about 10–15 years. For petroleum innovations, the standard deviation of this lag is about 5 years; in all other industries combined, it is about 16 years. Apparently, mechanical innovations require the shortest interval, and electronic innovations require the longest. The lag seems shorter for consumer products than for industrial products and shorter for innovations developed with government funds than for those developed with private funds.

---

time). These relatively few omissions are ignored, and we act as if we had the entire population of innovators in the analysis.

Of course, the assumption that $(e_j - \bar{e})$ and $(d_j - \bar{d})$ are statistically independent of $(t_j - t)/\Delta$ is rather bold. Some bias may result if d_j is higher immediately after the introduction of an innovation. If so, $(d_j - \bar{d})$ and $1 - (t_j - t)/\Delta$ may be negatively correlated, and we would probably overestimate $\bar{e}$ and underestimate $\bar{d}$. Where there were only a few innovators, this assumption (and the one in the previous paragraph) seemed particularly risky and we did not compute values of $\bar{e}$ and $\bar{d}$. But some preliminary work suggested that had we done so the results would have been much the same.

[33] If we had complete, year-by-year data on each firm's size, we could compute $\bar{e}$ and $\bar{d}$ without making the assumption discussed in note 32. The differences in growth rates shown in Table 6.7 are averages over periods of 1 to 10 years after an innovation was introduced. Obviously, the effects of an innovation decrease as time goes on. Finally, for the reason cited in note 32, the estimates of $\bar{d}$ may be biased downward. On the other hand, in the petroleum industry in 1947–1957, $\bar{d}$ may be unduly affected by one firm and is probably too high. Note, too, that the observed differences in growth rate may still be due in part to other factors that are associated with a firm's willingness to innovate and the timing of the innovation. Although the analysis in the text goes some way toward eliminating the problem, there is no way to be sure of its total elimination.

Second, in the steel and petroleum industries, process innovations were most likely to be introduced during periods when these industries were operating at about 75 percent of capacity: Contrary to the opinion of many economists, there was no tendency for innovations to cluster at the peak or the trough of the business cycle. Apparently, process innovation at the trough was discouraged by the meagerness of profits and the bleakness of future prospects; at the peak, it was discouraged by the lack of unutilized capacity. For product innovations, there was no evidence that the rate of innovation varied significantly over the business cycle.

Third, a simple investment function combining the flexible capacity accelerator with a simple model of innovation-induced investment can explain the behavior of investment in steel and petroleum somewhat better than the flexible accelerator alone. The timing of innovation seems to have had a significant effect on the level and timing of expenditures on plant and equipment. However, it is a difficult variable to handle empirically, and these results are presented merely as the findings of a crude experiment, not of a definitive study. The investment model that is considered is very rough.

Fourth, in the period immediately before they introduced the innovations, there was no persistent tendency for the successful innovators to grow more rapidly than other comparable firms. But in the period after they introduced the innovations, there was a considerable increase in the difference in growth rates between innovators and other comparable firms. In terms of short-term growth, the rewards for successful innovation seem to have been substantial, particularly for smaller firms.

Part IV

Diffusion of Innovations

CHAPTER 7

The Rate of Imitation

Previous chapters have been concerned with the invention of new techniques and products and with their initial introduction into practice. We have seen that there is often a long series of hurdles, technical and economic, that an invention must clear before it is introduced to the commercial market. Moreover, this is only part of the story. Once an invention is introduced for the first time, the battle is only partly won, since it must still gain widespread acceptance and use. The rate of diffusion is of great importance. The full social benefits of an innovation will not be realized if its use spreads too slowly.

Once a new technique is introduced by one firm, how soon do the others in the industry come to use it? What factors determine how rapidly they follow? The importance of these questions has long been recognized,[1] but until recently our knowledge of the imitation process did not extend far beyond Schumpeter's simple assertion that once a firm introduces a successful innovation, a host of imitators appear on the scene. This chapter summarizes some theoretical and empirical findings regarding the rate at which firms follow an innovator. A simple model is presented to help explain differences among innovations in the rate of imitation. The model is then tested against data showing how rapidly firms in four quite different industries came to use twelve innovations.

More specifically, the following section describes the twelve innovations and shows how rapidly firms followed the innovator in each case; in sections 2 to 4, we present and test a deterministic model constructed to explain the observed differences among these rates of imitation; a stochastic version is discussed in section 5; and some of the study's limitations are pointed out in section 6.

1. Rates of Imitation

This section describes how rapidly the use of twelve innovations spread from enterprise to enterprise in four industries—bituminous coal, iron and steel, brewing, and railroads. The innovations are the shuttle car, trackless mobile loader, and continuous mining machine (in bituminous coal); the by-product coke oven, continuous wide-strip mill, and continuous annealing line

[1] See Conference on Price Research, *Cost Behavior and Public Policy* (National Bureau of Economic Research, 1943).

for tin plate (in iron and steel); the pallet-loading machine, tin container, and high-speed bottle filler (in brewing); and the diesel locomotive, centralized traffic control, and car retarders (in railroads).

These innovations were chosen because of their outstanding importance and because it seemed likely that adequate data could be obtained for them. Excluding the tin container, all were types of heavy equipment permitting a substantial reduction in costs. The most recent of these innovations occurred after World War II; the earliest was introduced before 1900. In practically

FIGURE 7.1. Growth in the percentage of major firms that introduced twelve innovations, bituminous coal, iron and steel, brewing, and railroad industries, 1890–1958. (A). By-product coke oven (CO), diesel locomotive (DL), tin container (TC), and shuttle car (SC). (B). Car retarder (CR), trackless mobile loader (ML), continuous-mining machine (CM), and pallet-loading machine (PL). (C). Continuous wide-strip mill (SM), centralized traffic control (CTC), continuous annealing (CA), and high-speed bottle filler (BF).

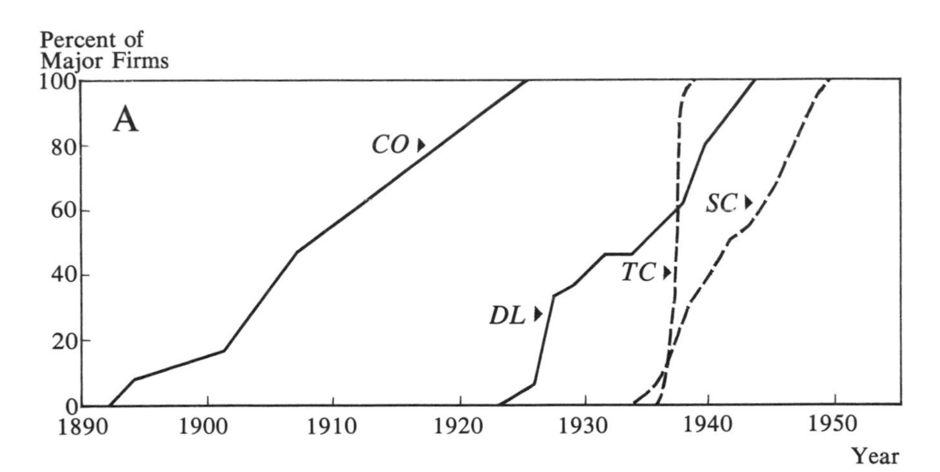

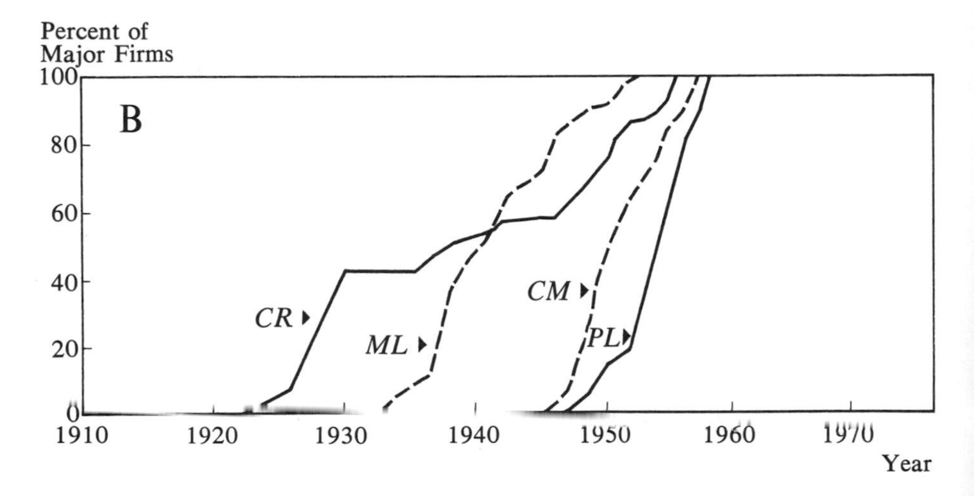

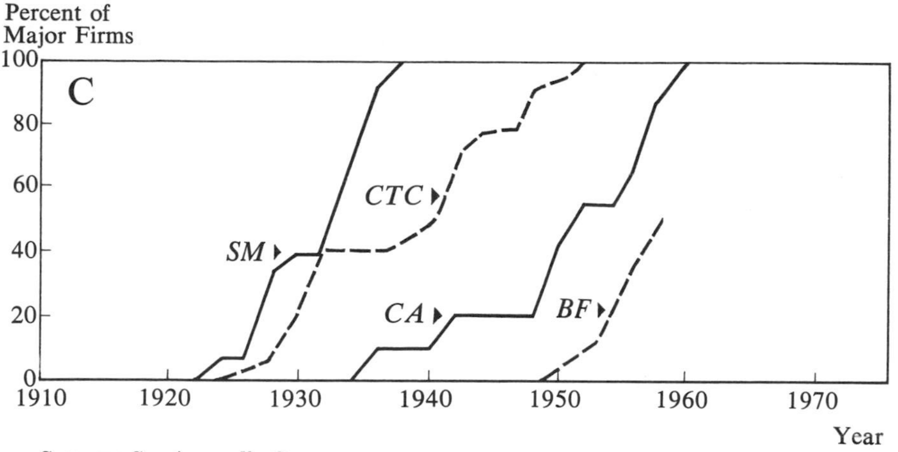

SOURCE: See Appendix C.

NOTE: For all but the by-product coke oven and tin container, the percentages given are for every 2 years from the year of initial introduction. Zero is arbitrarily set at 2 years prior to the initial introduction in these charts (but not in the analysis). The length of the interval for the by-product coke oven is about 6 years and for the tin container, it is 6 months. The innovations are grouped into the three sets shown above to make it easier to distinguish between the various growth curves.

all cases, the bulk of the development work was carried out by equipment manufacturers, and patents did not impede the imitation process.[2]

Figure 7.1 shows the percentage of major firms that had introduced each of these innovations at various points in time. To avoid misunderstanding, note three things regarding the data in Figure 7.1: (1) Because of difficulties in obtaining information concerning smaller firms and because in some cases they could not use the innovation in any event, only firms exceeding a certain size (given in Appendix C) are included.[3] (2) The percentage of firms having

[2] For descriptions of these innovations (and in some cases historical material), see Association of Iron and Steel Engineers, *The Modern Strip Mill*, for the strip mill; J. Camp and C. Francis, *The Making, Shaping, and Treating of Steel*, U.S. Steel Co., 1940, for continuous annealing and the by-product coke oven; annual issues of *Coal Age* (on mechanization), and American Mining Congress, Coal Mine Modernization Year Book (annual), for the innovations in coal; *Fortune* (Jan. 1936), for the tin container; *American Brewer*, Sept. 1954, for pallet loaders, and other issues for high-speed fillers; J. Jewkes, D. Sawers, and R. Stillerman, *The Sources of Invention* (New York: St. Martin's Press, 1958), for the diesel locomotive; E. Mansfield and H. Wein, "A Model for the Location of a Railroad Classification Yard," *Management Science* (Apr. 1958), pp. 292–293, for car retarders; and Union Switch and Signal Co., *Centralized Traffic Control*, Swissvale, 1931, for centralized traffic control. Actually, continuous annealing was often no cheaper than previous techniques, but it was required by customer demands. Its introduction was profitable since a firm's profits would have been lower without it.

[3] For the innovations in the steel industry, we imposed the particular size limits cited in Appendix C because, according to interviews, it seemed very unlikely that firms smaller than this would have been able to use them. For the innovations in the coal and brewing industries, there were no adequate published data concerning the dates when particular firms first introduced them, and we had to get the information directly from the firms. It

introduced an innovation, regardless of the scale on which they did so, is given. The possible objections to this are largely removed by the fact that these innovations had to be introduced on a fairly large scale. By using them at all, firms made a relatively heavy financial commitment.[4] (3) In a given industry, most of the firms included for one innovation are also included for the others. Thus the data for each of the innovations are quite comparable in this regard.

Two conclusions regarding the rate of imitation emerge from Figure 7.1. First, the diffusion of a new technique is generally a rather slow process. Measuring from the date of the first successful commercial application,[5] it took 20 years or more for all the major firms to install centralized traffic control, car retarders, by-product coke ovens, and continuous annealing. Only in the case of the pallet-loading machine, tin container, and continuous mining machine did it take 10 years or less for all the major firms to install them.

Second, the rate of imitation varies widely. Although it sometimes took decades for firms to install a new technique, in other cases they followed the innovator very quickly. For example, it took about 15 years for half of the major pig-iron producers to use the by-product coke oven, but only about 3 years for half of the major coal producers to use the continuous mining machine. The number of years elapsing before half the firms had introduced an innovation varied from 0.9 to 15, the average being 7.8.

2. A Deterministic Model

Why were these firms so slow to install some innovations and so quick to install others? What factors seem to govern the rate of imitation? In this section, we construct a simple deterministic model to explain the results shown in Figure 7.1. In sections 3 and 4, we test this model and see the effects of introducing some additional variables into it.

The following notation is used. Let n_{ij} be the total number of firms on which the results in Figure 7.1 for the jth innovation in the ith industry are

---

seemed likely that "nonresponse" would be a considerable problem among the smaller firms. This consideration, as well as the fact that the smallest firms often could not use them, led us to impose the rather arbitrary lower limits on size shown in Appendix C. For the railroad innovations, we took firms large enough to use car retarders and centralized traffic control and, to insure comparability, the same size limits were used for the diesel locomotive.

[4] The only alternative would be to take the date when a firm first used the innovation to produce some specified percentage of its output. In almost every case, such data were not published and it would have been extremely difficult, if not impossible, to obtain them from the firms. To install a strip mill, by-product coke ovens, continuous annealing, or car retarders, a firm had to invest many millions of dollars. Even for shuttle cars, trackless mobile loaders, canning equipment, and pallet loaders, the investment (although less than $100,000 usually) was by no means trivial in these industries.

[5] Note that we measure how quickly other firms followed the one that first *successfully* applied the technique. Others may have tried roughly similar things before but failed. By a successful application, we mean one where the equipment was used commercially for years, not installed and quickly withdrawn.

based[6] ($j = 1,2,3$; $i = 1,2,3,4$). Let $m_{ij}(t)$ be the number of these firms having introduced this innovation at time t, π_{ij} be the profitability of installing this innovation relative to that of alternative investments, and S_{ij} be the investment required to install this innovation as a percent of the average total assets of these firms. More precise definitions of S_{ij} and π_{ij} are provided in section 3. Let $\lambda_{ij}(t)$ be the proportion of "hold-outs" (firms not using this innovation) at time t that introduce it by time $t + 1$, that is,

$$\lambda_{ij}(t) = \frac{m_{ij}(t+1) - m_{ij}(t)}{n_{ij} - m_{ij}(t)}. \tag{1}$$

Our basic hypothesis can be stated quite simply. We assume that the proportion of "hold-outs" at time t that introduce the innovation by time $t + 1$ is a function of (1) the proportion of firms that already introduced it by time t, (2) the profitability of installing it, (3) the size of the investment required to install it, and (4) other unspecified variables. Allowing the function to vary among industries, we have

$$\lambda_{ij}(t) = f_i\left(\frac{m_{ij}(t)}{n_{ij}}, \pi_{ij}, S_{ij}, \ldots\right). \tag{2}$$

In the following few paragraphs, we take up the presumed effects of variation in $\frac{m_{ij}(t)}{n_{ij}}$, π_{ij}, and S_{ij} on $\lambda_{ij}(t)$ and the reasons for interindustry differences in the function.

First, one would expect that increases in the proportion of firms already using an innovation would increase $\lambda_{ij}(t)$. As more information and experience accumulate, it becomes less of a risk to begin using it.[7] Competitive pressures mount and "bandwagon" effects occur. Where the profitability of

[6] That is, the total number of firms for which we have data. See Table 7.1 for the n_{ij}.

[7] The profitability of installing each of these innovations was viewed at first with considerable uncertainty. For example, there was great uncertainty about maintenance costs for diesel locomotives, "down-time" for continuous mining machines, the safety of centralized traffic control, and the useful life of by-product coke ovens. The perceived risks seldom disappeared after only a few firms had introduced them; in some cases according to interviews, it took many years. This helps to account for the fact (noted above) that the imitation process generally went on rather slowly. This also suggests the potential importance of two variables not recognized explicitly in equation (2): first, the extent of the initial uncertainty concerning the profitability of an innovation (depending in part on the extent of prior field testing by manufacturers), and second, the rate at which this uncertainty declined. For some types of innovations, a few installations and a relatively short period of use can cut the risks to little or nothing. For other types, installations under many sorts of conditions and a long period of use (to be sure of useful life, maintenance costs, etc.) are required. See section 6 for further discussion. Note that all these innovations are new processes or (in the case of the tin container) a packaging innovation. This model would not be applicable to some innovations, like an entirely new product, where, as more firms produce it, it becomes less profitable for others to do so. There is no evidence of significant decreases of this sort in these cases. (For example, as more firms introduced diesel locomotives, this did not make it less profitable for others to do so.) Moreover, the model presumes that π_{ij} is appreciably greater than unity (which is the case here). Finally, the data support the hypothesis in the text (see note 8).

using the innovation is very difficult to estimate, the mere fact that a large proportion of its competitors have introduced it may prompt a firm to consider it more favorably. Both interviews with executives in the four industries and the data in Figure 7.1 indicate that this is the case.[8]

Second, the profitability of installing the innovation would also be expected to have an important influence on $\lambda_{ij}(t)$. The more profitable this investment is relative to others that are available, the greater is the chance that a firm's estimate of the profitability will be high enough to compensate for whatever risks are involved and that it will seem worthwhile to install the new technique rather than to wait. As the difference between the profitability of this investment and that of others widens, firms tend to note and respond to the difference more quickly. Both the interviews and the few other studies regarding the rate of imitation suggest that this is so.[9]

Third, for equally profitable innovations, $\lambda_{ij}(t)$ should tend to be smaller for those requiring relatively large investments. One would expect this on the grounds that firms tend to be more cautious before committing themselves to such projects and that they often have more difficulty in financing them. According to the interviews, this factor is often important.

Finally, for equally profitable innovations requiring the same investment, $\lambda_{ij}(t)$ is likely to vary among industries. It might be higher in one industry than in another because firms in the former industry have less aversion to risk, because markets are more keenly competitive, because the attitude of the labor force toward innovation is more favorable, or because the industry is healthier financially. Casual observation suggests that such interindustry differences may have a significant effect on $\lambda_{ij}(t)$.

Returning to equation (2), we act as though the number of firms having introduced an innovation can vary continuously rather than assume only

[8] Beginning with the date when $m_{ij}(t) = 1$, we computed $\lambda_{ij}(t)$ and $m_{ij}(t)/n_{ij}$, using the intervals in the note to Figure 7.1 as time units and stopping when $m_{ij}(t) = n_{ij}$. Then we calculated the correlation between $\lambda_{ij}(t)$ and $m_{ij}(t)/n_{ij}$. The correlation coefficients were 0.77 (continuous mining machine), 0.85 (by-product coke oven), 0.49 (tin container), 0.65 (centralized traffic control), 0.85 (continuous wide-strip mill), 0.66 (diesel locomotive), 0.52 (car retarders), 0.55 (bottle fillers), 0.96 (pallet loaders), 0.81 (continuous annealing), 0.46 (trackless mobile loaders), and 0.94 (shuttle car). Using a one-tailed test (which is appropriate here), all coefficients but those for the continuous mining machine, bottle filler, and trackless mobile loader are significant (0.05 level). All are positive.

Of course, others have stated similar hypotheses before. For example, Schumpeter noted that "accumulating experience and vanishing obstacles" smooth the way for imitators (J. Schumpeter, *Business Cycles* [New York: McGraw-Hill, 1939]), and Coleman, Katz, and Menzel noted a "snow-ball" effect (J. Coleman, E. Katz, and H. Menzel, "The Diffusion of an Innovation Among Physicians," *Sociometry* [Dec. 1957]). For what it is worth, almost all the executives we interviewed considered this effect to be present. The number of installations of the innovation or the number of firms using it might have been used rather than the proportion of firms, but the latter seems to work quite well.

[9] See Z. Griliches, "Hybrid Corn: An Exploration in the Economics of Technological Change," *Econometrica* (Oct. 1957). Another study of the rate of imitation is found in J. Yance, "Technological Change as a Learning Process: The Dieselization of the Railroads" (unpublished, 1957). Both papers focus on only one innovation (hybrid corn and the diesel locomotive). Of course, the interfirm variation in profitability, as well as the average, could influence $\lambda_{ij}(t)$.

integer values, and we assume that $\lambda_{ij}(t)$ can be approximated adequately within the relevant range by a Taylor's expansion that drops third- and higher-order terms. Assuming that the coefficient of $(m_{ij}(t)/n_{ij})^2$ in this expansion is zero (and the data in Figure 7.1 generally support this),[10] we have

(3)
$$\lambda_{ij}(t) = a_{i1} + a_{i2}\frac{m_{ij}(t)}{n_{ij}} + a_{i3}\pi_{ij} + a_{i4}S_{ij} + a_{i5}\pi_{ij}\frac{m_{ij}(t)}{n_{ij}} + a_{i6}S_{ij}\frac{m_{ij}(t)}{n_{ij}} + a_{i7}\pi_{ij}S_{ij} + a_{i8}\pi_{ij}^2 + a_{i9}S_{ij}^2 + \cdots$$

where additional terms contain the unspecified variables in equation (2). Thus

(4)
$$m_{ij}(t+1) - m_{ij}(t) = [n_{ij} - m_{ij}(t)]\left[a_{i1} + a_{i2}\frac{m_{ij}(t)}{n_{ij}} + \cdots + a_{i9}S_{ij}^2 + \cdots\right].$$

Assuming that time is measured in fairly small units, we can use as an approximation the corresponding differential equation[11]

(5)
$$\frac{dm_{ij}(t)}{dt} = [n_{ij} - m_{ij}(t)]\left[Q_{ij} + \phi_{ij}\frac{m_{ij}(t)}{n_{ij}}\right],$$

the solution of which is

(6)
$$m_{ij}(t) = \frac{n_{ij}\left[e^{l_{ij}+(Q_{ij}+\phi_{ij})t} - \dfrac{Q_{ij}}{\phi_{ij}}\right]}{1 + e^{l_{ij}+(Q_{ij}+\phi_{ij})t}},$$

where l_{ij} is a constant of integration, Q_{ij} is the sum of all terms in equation (3) not containing $m_{ij}(t)/n_{ij}$, and

(7)
$$\phi_{ij} = a_{i2} + a_{i5}\pi_{ij} + a_{i6}S_{ij} + \cdots.$$

Of course, ϕ_{ij} is the coefficient of $m_{ij}(t)/n_{ij}$ in equation (3).

To get any further, we must impose additional constraints on the way $m_{ij}(t)$ can vary over time. One simple condition we can impose is that as we go backward in time, the number of firms having introduced the innovation must tend to zero,[12] that is,

[10] To test this assumption for these innovations, we used $(m_{ij}(t)/n_{ij})^2$ as an additional independent variable in the regression described in note 8 and used the customary analysis of variance to determine whether this resulted in a significant increase in the explained variation. For all but continuous annealing, car retarders, and the diesel locomotive, it does not (and in these cases, the increase is often barely significant). Hence, in most cases, there is no evidence that this coefficient is nonzero.

[11] This is like some approximations commonly used in capital theory. For example, one often replaces equations like $x(t+1) - x(t) = rx(t)$ with $\dot{x}(t) = rx(t)$.

[12] Of course, other conditions could be imposed in addition or instead. For example, in section 5, we take as given the date when a particular number of firms had installed an innovation and force $m_{ij}(t)$ to equal that number at that date. But equation (8) is all we need for present purposes. Note that it implies that Q_{ij} is zero. The data described in note 8 are

(8) $$\lim_{t \to -\infty} m_{ij}(t) = 0.$$

Using this condition, it follows that

(9) $$m_{ij}(t) = n_{ij}[1 + e^{-(l_{ij}+\phi_{ij}t)}]^{-1}.$$

Thus, the growth over time in the number of firms having introduced an innovation should conform to a logistic function, an S-shaped growth curve frequently encountered in biology and the social sciences.[13]

If equation (9) is correct, it can be shown that the rate of imitation is governed by only one parameter—ϕ_{ij}.[14] If we assume that the sum of the unspecified terms in equation (7) is uncorrelated with π_{ij} and S_{ij} and that it can be treated as a random error term,

(10) $$\phi_{ij} = b_i + a_{i5}\pi_{ij} + a_{i6}S_{ij} + z_{ij},$$

where b_i equals a_{i2} plus the expected value of this sum and z_{ij} is a random variable with zero expected value. Hence, the expected value of ϕ_{ij} in a particular industry is a linear function of π_{ij} and S_{ij}.

To sum up, the model leads to the following two predictions. First, the number of firms having introduced an innovation, if plotted against time, should approximate a logistic function. Second, the rate of imitation in a particular industry should be higher for more profitable innovations and innovations requiring relatively small investments. More precisely, ϕ_{ij}, a measure of the rate of imitation, should be linearly related to π_{ij} and S_{ij}.

3. Tests of the Model

We test this model in two steps: (1) by estimating ϕ_{ij} and l_{ij} and determining how well equation (9) fits the data, and (2) by seeing whether the

consistent with this, but even if Q_{ij} were nonzero but small (and it certainly could not be large), equation (9) should be a reasonably good approximation. Note too that, if the model holds, $\phi_{ij} > 0$.

[13] Note that things are simplified here by the fact that all firms we consider eventually introduced these innovations. (A few went out of business first, but not because of the appearance of the innovation. See Appendix C.) Had this not been the case, it would have been necessary either to provide a mechanism explaining the proportion that did not do so or to take it as given. Of course, by taking only the larger firms, we made sure that all could use these innovations. The smaller firms (that were not potential users) are omitted. For the high-speed bottle filler, we make the reasonable assumption that all the major firms will ultimately introduce it. Note that the argument here is different from that generally used in biology to arrive at the logistic function and that it explicitly includes variables affecting its shape.

[14] It seems reasonable to take, as a measure of the rate of imitation, the time span between the date when 20 percent (for example) of the firms had introduced an innovation, and the date when 80 percent (for example) had done so. According to the model, this time span equals 2.77 ϕ_{ij}^{-1} and is therefore independent of l_{ij}. Of course, 20 and 80 are arbitrary choices. (For example, 10 and 90 are used in Chapter 9.) If, rather than 20 and 80, we take P_1 and P_2, it can be shown that the time span equals $\phi_{ij}^{-1}\, ln\, [(1 - P_1)P_2/P_1(1 - P_2)]$. The important point is that it depends only on ϕ_{ij}, not l_{ij}, regardless of which values of P_1 and P_2 we choose.

expected value of ϕ_{ij} seems to be a linear function of π_{ij} and S_{ij}. The results of these tests suggest that the model can explain the results in Figure 7.1 quite well.

To carry out the first step, note that, if the model is correct, it follows from equation (9) that

$$(11) \qquad ln\left[\frac{m_{ij}(t)}{n_{ij} - m_{ij}(t)}\right] = l_{ij} + \phi_{ij}t.$$

Measuring time in years (from 1900), treating this as a regression equation, and using least-squares (after properly weighting the observations [13]), we derive estimates of l_{ij} and ϕ_{ij} (Table 7.1).[15] To see how well equation (9) can represent the data, we insert these estimates into equation (9) and compare the calculated increase over time in the number of firms having introduced each innovation with the actual increase.

Judging merely by a visual comparison, the calculated growth curves generally provide reasonably good approximations to the actual ones. However, the "fit" is not uniformly good. For the railroad innovations, there is evidence of serial correlation among the residuals, and the approximations are less satisfactory than for the other innovations. Table 7.1 contains two rough measures of "goodness-of-fit": The root-mean-square deviation of the actual from the computed number of firms having introduced the innovation[16] and the coefficient of correlation between $ln\ [m_{ij}(t)/n_{ij} - m_{ij}(t)]$ and t.[17] They seem to bear out the general impression that equation (9) represents the data for most of the innovations quite well.[18]

[15] When $ln\ [m_{ij}(t)/n_{ij} - m_{ij}(t)]$ was infinite, the observation was omitted. The observations were 1 year apart for the continuous wide-strip mill, continuous mining machine, and centralized traffic control, 6 years apart for the by-product coke oven, 1 month apart for the tin container, and 2 years apart for the others. For the high-speed bottle filler, we used all of the data available when this study was performed, but the imitation process was not complete at that time.

[16] For each innovation, the difference between the computed and actual values of $m_{ij}(t)$ was obtained for $t = t^*_{ij}, t^*_{ij} + 1, \ldots, t^{**}_{ij}$, where time is measured in the units described in note 15, t^*_{ij} is the first date when $m_{ij}(t) = 1$, and t^{**}_{ij} is the first date when $m_{ij}(t) = n_{ij}$. Then the root-mean-square of these differences was obtained. Note that this is a measure of absolute, not relative, error, and when comparing the results for different innovations, take account of differences in n_{ij}.

[17] This coefficient is labeled r_{ij} in Table 7.1. Note that it is based on weighted observations (the weights being those suggested by J. Berkson, in "A Statistically Precise and Relatively Simple Method of Estimating the Bio-Assay with Quantal Response, Based on the Logistic Function," *Journal of the American Statistical Association* (Sept. 1953)).

[18] Of course, the logistic function is not the only one that might represent the data fairly adequately. (For example, the normal cumulative distribution function would probably do as well.) And, since the actual curve has to be roughly S-shaped, it is not surprising that it fits reasonably well. There seemed to be little point in attempting any formal "goodness-of-fit" tests here. The number of firms considered in each case is quite small, and the tests would not be very powerful. All that we conclude from Table 7.1 is that equation (9) provides a reasonably adequate description of the data and hence that ϕ_{ij} is generally reliable as a measure of the rate of imitation.

Table 7.1. *Parameters, Estimates, and Root-Mean-Square Errors: Deterministic and Stochastic Models*

INNOVATION	PARAMETER						
	n_{ij}	π_{ij}	S_{ij}	d_{ij}	g_{ij}	t^*_{ij}	δ_{ij}
Diesel locomotive	25	1.59	.015	35	1.00	1925	1
Centralized traffic control	24	1.48	.024	0	1.50	1926	1
Car retarders	25	1.25	.785	0	1.50	1924	0
Continuous wide-strip mill	12	1.87	4.908	30	4.50	1924	0
By-product coke oven	12	1.47	2.083	10	4.00	1894	1
Continuous annealing	9	1.25	.554	9	4.25	1936	1
Shuttle car	15	1.74	.013	9	1.25	1937	0
Trackless mobile loader	15	1.65	.019	6	2.50	1934	1
Continuous mining machine	17	2.00	.301	8	1.00	1947	1
Tin container	22	5.07	.267	0	6.50	1935	1
High-speed bottle filler	16	1.20	.575	10	2.25	1951	1
Pallet-loading machine	19	1.67	.115	0	2.25	1948	1

	ESTIMATES AND ROOT-MEAN-SQUARE ERRORS					
	$\hat{l}_{ij}$	$\hat{\phi}_{ij}$	r_{ij}	$\hat{\theta}_{ij}$	ERROR (Det.)	ERROR (Stoch.)
Diesel locomotive	−6.64	0.20	0.89	0.30	2.13	5.63
Centralized traffic control	−7.13	.19	.94	.24	1.52	3.44
Car retarders	−3.95	.11	.90	.17	2.08	5.02
Continuous wide-strip mill	−10.47	.34	.95	.42	.83	.90
By-product coke oven	−1.47	.17	.98	.18	.16	.84
Continuous annealing	−8.51	.17	.93	.22	.74	1.42
Shuttle car	−13.48	.32	.95	.45	.86	2.03
Trackless mobile loader	−13.03	.32	.97	.39	.71	1.66
Continuous mining machine	−24.96	.49	.98	.59	.81	2.22
Tin container	−84.35	2.40	.96	2.64	1.34	3.00
High-speed bottle filler	−20.58	.36	.97	.40	.56	.95
Pallet-loading machine	−29.07	.55	.97	.63	1.10	1.58

SOURCE: See Appendix C and notes 6, 15, 16, 17, 20, 22, 25, 29, 32, and 34.

To carry out the second step, we assume that a_{i5} and a_{i6} do not vary among industries.[19] Thus equation (10) becomes

(12) $$\phi_{ij} = b_i + a_5\pi_{ij} + a_6S_{ij} + z_{ij},$$

[19] Because of the small number of observations, we are forced to make this assumption. If data were available for more innovations, it would not be necessary. Of course, if inter-industry differences in these coefficients are statistically independent of π_{ij} and S_{ij}, a_5 and a_6 can be regarded as averages and there is no real trouble.

and, if we assume that errors in the estimates of ϕ_{ij} are uncorrelated with π_{ij} and S_{ij}, we have

$$\hat{\phi}_{ij} = b_i + a_5\pi_{ij} + a_6 S_{ij} + z'_{ij}, \tag{13}$$

where $\hat{\phi}_{ij}$ is the estimate of ϕ_{ij} derived above. Assuming that z'_{ij} is distributed normally with constant variance, standard tests can be applied to determine whether a_5 and a_6 are nonzero.

Before discussing the results of these tests, we must describe how π_{ij} and S_{ij} are measured. We obtained from as many of the firms as possible estimates of the pay-out period for the initial installation of the innovation and the pay-out period required from investments. These estimates were derived primarily from correspondence, but published materials were used when possible.[20] Then the average pay-out period required by the firms (during the relevant period) to justify investments divided by the average pay-out period for the innovation was used as a measure of π_{ij}.[21] To measure S_{ij}, we used the average initial investment in the innovation as a percentage of the average

[20] A letter was sent to each firm asking how long it took for the initial investment in the innovation to pay for itself and what the required pay-out period was during the relevant period. (For a definition of the pay-out period, see F. Swalm, "On Calculating the Rate of Return of an Investment," *Journal of Industrial Engineering* [Mar. 1958], or any standard text on capital budgeting. Both the realized and required pay-out periods are before taxes.) Replies were received from 50 percent of the railroads and coal producers and 20 percent of the steel companies and breweries. The estimates for the railroad innovations are probably most accurate since the firms referred us to fairly reliable published studies (see Appendix C). The data for the coal innovations are probably quite good too. The estimates for the steel innovations (particularly the by-product coke oven and continuous annealing) and the tin container are probably least accurate because they occurred so long ago and because fewer firms provided data. Despite the fact that the averages are probably more accurate than the individual figures and that the results were checked with other sources in interviews (see Appendix C), the estimates of the π_{ij} are rough. However, if we omit the cases where the estimates are poorest, the results seem to be largely unaffected.

[21] For relatively long-lived investments, the reciprocal of the pay-out period is a fairly adequate approximation to the rate of return. See M. Gordon "The Payoff Period and the Rate of Profit," *Journal of Business* (Oct. 1955) and Swalm, "On Calculating the Rate of Return." Hence this measure of π_{ij} is approximately equal to the average rate of return derived (ex post) from the innovation divided by the average rate of return firms required (ex ante) to justify investments. Of course, it would be more appropriate for firms to recognize that introducing it next year, the year after, etc., are the alternatives to introducing it now and to make an analysis like G. Terborgh's, in *Dynamic Equipment Policy* (New York: McGraw-Hill, 1949). But as he points out, most firms seem to make these decisions on the basis of the "rate of return" or "pay-out period"; and thus it seems reasonable to use such measures in a study (like this) where we try to explain behavior, not prescribe it. Even so, this measure is only an approximation. The rate of return from the investment in an innovation is measured ex post, not ex ante. The average rate of return that could have been realized by the "hold-outs" at a particular point in time probably differed from the average rate of return actually realized by all firms. Our estimates are based implicitly on the factor prices and age of old equipment (where replacement occurred) that prevailed when the innovation was installed, and these (and other) factors, as well as the innovation itself, do not remain unchanged. According to the interviews, the average return that could have been realized probably varied over time about an average that was highly correlated with our estimate, and hence the latter provides a fair indication of the level. But the parameters describing the temporal variation about this level are among the "other" variables in equation (2). See section 6 and note 40.

total assets of the firms (during the relevant period).[22] Table 7.1 contains the results.

Using these rather crude data, we obtained least-squares estimates of the parameters in equation (13) and tested whether they were zero. The resulting equation is

$$\textbf{(14)} \qquad \hat{\phi}_{ij} = \begin{Bmatrix} -0.29 \\ -0.57 \\ -0.52 \\ -0.59 \end{Bmatrix} + \underset{(0.015)}{0.530\pi_{ij}} - \underset{(0.014)}{0.027S_{ij}}, \quad (r = 0.997)$$

in which the top figure in the braces pertains to the brewing industry, the next to coal, the following to steel, and the bottom figure pertains to the railroads. The coefficients of π_{ij} and S_{ij} have the expected signs (indicating that increases in π_{ij} and decreases in S_{ij} increase the rate of imitation), and both differ significantly from zero.

Chapter 5 studied the effects of market structure on the lag between invention and innovation. A related, and equally important, topic is the effect of market structure on the rate of diffusion. When π_{ij} and S_{ij} are held constant, equation (14) shows that there are significant interindustry differences in $\hat{\phi}_{ij}$, the rate of imitation being particularly high in brewing. These differences seem to be broadly consistent with the hypothesis often advanced that the rate of imitation is higher in more competitive industries, but there are too few data to warrant any real conclusion on this score.[23]

The scatter diagram in Figure 7.2 shows that equation (14) represents the data surprisingly well. When corrected for the relatively few degrees of freedom, the correlation coefficient is 0.997. Of course, one point (the tin container) strongly affects the results. But if that point is omitted, the interindustry differences remain much the same, the coefficients of π_{ij} and S_{ij}

[22] For the steel, coal, and brewing industries, we obtained the total assets of as many of these firms as possible (during the relevant period) from *Moody's*. For the railroads, the 1936 reproduction costs in L. Klein, "Studies in Investment Behavior," *Conference on Business Cycles* (National Bureau of Economic Research, 1951) were used for the firms. Estimates of the approximate investment required to install each innovation were obtained primarily from the interviews (described in Appendix C), and each was divided by the average total assets of the relevant firms. Since the investment could vary considerably, the results are only approximate.

[23] For a more extended statement of this hypothesis, see J. Robinson, *The Accumulation of Capital* (Homewood, Ill.: Irwin, 1956). We encounter here the usual difficulty in measuring the "extent of competition." But most economists would undoubtedly agree that brewing and coal are more competitive than steel and railroads (and the average value of b_i is larger in the former industries). When six of my colleagues were asked to rank them by "competitiveness," they put brewing first, coal second, iron and steel third, and railroads fourth. If these ranks are correlated with the estimates of b_i in equation (14), the rank correlation coefficient is positive (0.80), but not statistically significant. (The 0.05 probability level is used throughout this chapter.) There are too few industries to allow a reasonably powerful test of this hypothesis even if our procedures were refined somewhat, but for most rankings that seem sensible, the correlation in the sample is positive.

FIGURE 7.2. Plot of actual $\hat{\phi}_{ij}$ against that computed from equation (14), twelve innovations.

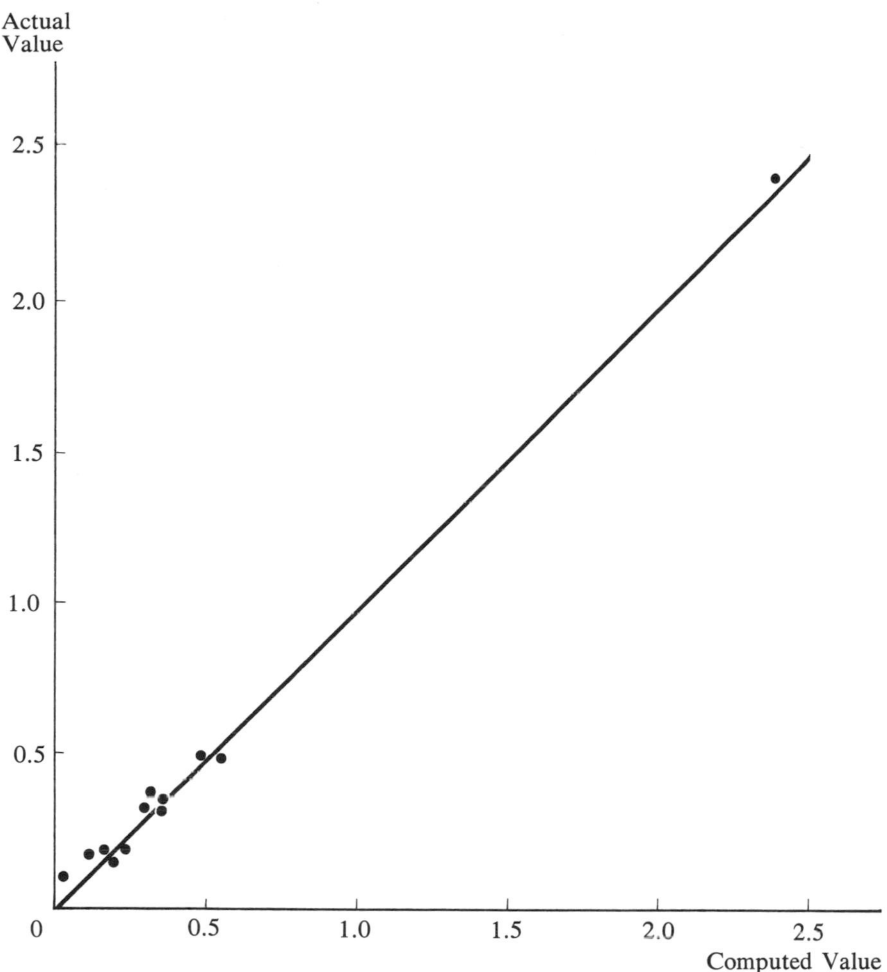

SOURCE: Table 7.1 and equation (14).

NOTE: The difference between an actual and computed value of $\hat{\phi}_{ij}$ is equal to the vertical difference between the point and the 45-degree line.

keep the same signs (the latter becoming nonsignificant), and the correlation coefficient, corrected for degrees of freedom, is still 0.97.[24]

[24] Five points should be noted: First, the tin container is a somewhat different type of innovation from the others, and the profitability data for it are probably least reliable (see note 20). Thus, besides its being an extreme point in Figure 7.2, there are other possible grounds for excluding it, and it is reassuring to note that its exclusion has so little effect on the results. Second, because of the way it is constructed, the model can only be expected to work if π_{ij} and S_{ij} remain within certain bounds. If π_{ij} is close to one and/or S_{ij} is very large, it is likely to perform poorly. Third, one-tailed tests are used to determine the significance of the coefficients of π_{ij} and S_{ij}. Fourth, it is noteworthy that $\hat{\phi}_{ij}$, not a very obvious

Hence, the model seems to fit the data quite well. In general, the growth in the number of users of an innovation can be approximated by a logistic curve. And there is definite evidence that more profitable innovations and innovations requiring smaller investments had higher rates of imitation, the particular relationship being strikingly similar to the one we predicted. Though it is no more than a simple first approximation, the model can represent the empirical results shown in Figure 7.1 surprisingly well.

4. Additional Factors

Of course, other factors may also have been important, and their inclusion in the model may permit a significantly better explanation of the differences among rates of imitation. Moreover, it may show that the apparent effect of the previously mentioned variables on the rate of imitation is partly due to their influence. We turn now to a discussion of the influence of four other factors on $\lambda_{ij}(t)$.

First, one might expect $\lambda_{ij}(t)$ to be smaller if the innovation replaces equipment that is very durable. In such cases there is a good chance that a firm's old equipment still has a relatively long useful life, according to past estimates. Although rational economic calculation might indicate that replacement would be profitable, firms may be reluctant to scrap equipment that is not fully written off and that will continue to serve for many years. If so, d_{ij}—the number of years that typically elapsed before the old equipment was replaced (before the innovation appeared)—may be one of the excluded variables in equation (2), and hence $\hat{\phi}_{ij}$ may be a linear function of π_{ij}, S_{ij}, and d_{ij}.[25] If d_{ij} is included in equation (14),

$$(15) \quad \hat{\phi}_{ij} = \begin{Bmatrix} -0.28 \\ -0.56 \\ -0.51 \\ -0.57 \end{Bmatrix} + \underset{(0.015)}{0.528\pi_{ij}} - \underset{(0.015)}{0.020S_{ij}} - \underset{(0.0014)}{0.0017d_{ij}}. \quad (r = 0.997)$$

measure of the rate of imitation, should be linearly related to π_{ij} and S_{ij}, as the model predicts. When the result in note 14 is used, the number of years elapsing between when 20 percent and when 80 percent introduced the innovation can easily be used instead as the dependent variable in equation (14). The resulting relationship is nonlinear. Fifth, this model may also help to explain the empirical results Griliches obtained in his excellent study of the regional acceptance of hybrid corn (in "Hybrid Corn").

[25] We carry d_{ij} from equation (2) on and make the same assumptions we did with regard to π_{ij} and S_{ij}, and it follows that ϕ_{ij} should be a linear function of π_{ij}, S_{ij}, and d_{ij}. This of course is also true for the other variables discussed in this section. Estimates of d_{ij} were obtained primarily from the interviews. If an innovation was largely a supplement or addition to old plant (like centralized traffic control and car retarders), if it served a different purpose than the old equipment (like canning equipment vs. bottling equipment), and if it displaced only labor (like pallet loaders), we let d_{ij} equal zero. In such cases, there appeared to be no equipment whose durability could influence the decision significantly. Note that the age distribution and number of units of old equipment are also important in determining how long it takes before the first one "wears out." It would also have been preferable to have included the profitability of replacing old equipment of various ages rather than just the average figure used. But the available data would not permit this. For further discussion, see Chaps. 8 and 9.

Though there is some apparent tendency for the rate of imitation to be lower in cases in which very durable equipment had to be replaced, it is not statistically significant. (For the values of d_{ij} and the other factors discussed below, see Table 7.1.)

Second, one might expect $\lambda_{ij}(t)$ to be higher if firms are expanding at a rapid rate. If they are convinced of its superiority, the innovation will be introduced in the new plants built to accommodate the growth in the market. If there is little or no expansion, its introduction must often wait until the firms decide to replace existing equipment. If g_{ij}—the annual rate of growth of industry sales during the period—is included in equation (14),

$$\textbf{(16)} \quad \hat{\phi}_{ij} = \begin{Bmatrix} -0.32 \\ -0.56 \\ -0.53 \\ -0.58 \end{Bmatrix} + \underset{(0.032)}{0.484\pi_{ij}} - \underset{(0.013)}{0.025S_{ij}} + \underset{(0.026)}{0.042g_{ij}}. \quad (r = 0.998)$$

Thus, there is some apparent tendency for the rate of imitation to be higher where output was expanding at a very rapid rate, but it is not statistically significant.[26]

Third, $\lambda_{ij}(t)$ may have increased over time. This hypothesis has been advanced by economists on numerous occasions.[27] Presumably, the reasons for such a trend would be the evolution of better communication channels, more sophisticated methods to evaluate machine replacement, and more favorable attitudes toward frequent changes in techniques. If t_{ij}—the year (less 1900) when the innovation was first introduced[28]—is included in equation (14),

$$\textbf{(17)} \quad \hat{\phi}_{ij} = \begin{Bmatrix} -0.37 \\ -0.64 \\ -0.55 \\ -0.63 \end{Bmatrix} + \underset{(0.016)}{0.535\pi_{ij}} - \underset{(0.015)}{0.027S_{ij}} + \underset{(0.0014)}{0.0014t_{ij}}. \quad (r = 0.997)$$

There is some apparent tendency for the rate of imitation to increase over time, but it is not statistically significant.

[26] Of course, the effect of g_{ij} may depend on whether old equipment must be replaced, how durable it is, the difference between the profitability of replacement and of installing the innovation in new plant, the extent of excess capacity at the beginning of the period, the size of a plant relative to the size of the market, etc. (For some relevant discussion, see T. Scitovsky, "Economies of Scale and European Integration," *American Economic Review* [Mar. 1956].) The effect of this factor, like d_{ij}, reflects the possible unwillingness of firms to scrap existing equipment. It would have been preferable to have used figures on the profitability of using the innovation in new plant as well as to replace equipment of various ages, but data were not available. Note too that g_{ij} may be affected by the appearance of the innovation. For a description of the data on g_{ij}, see Appendix C. If S_{ij} is omitted from equations (15) to (18), the coefficients of d_{ij}, g_{ij}, etc. remain nonsignificant.

[27] For example, see R. Mack, *The Flow of Business Funds and Consumer Purchasing Power* (Columbia, 1941), p. 295; and H. Jerome, *Mechanization in Industry* (National Bureau of Economic Research, 1934), p. xxv.

[28] That is, $t_{ij} = t^*_{ij} - 1900$. See note 32.

Finally, one might suppose that $\lambda_{ij}(t)$ would be influenced by the phase of the business cycle during which the innovation was first introduced. Let δ_{ij} equal one if the innovation is introduced in the expansion phase and zero if it is introduced in the contraction phase.[29] Including δ_{ij} in equation (14), we have

$$\text{(18)} \quad \hat{\phi}_{ij} = \begin{Bmatrix} -0.26 \\ -0.55 \\ -0.48 \\ -0.57 \end{Bmatrix} + \underset{(0.017)}{0.530\pi_{ij}} - \underset{(0.020)}{0.033S_{ij}} - \underset{(0.045)}{0.022\delta_{ij}}, \quad (r = 0.997)$$

and the effect of δ_{ij} turns out to be nonsignificant.

Thus, our results regarding these additional factors are largely inconclusive. Though each might be expected to have some effect, their inclusion in the analysis does not lead to a significantly better explanation of the observed differences in $\hat{\phi}_{ij}$. Their apparent effects are in the expected direction, but data for more innovations will be required before one can be reasonably sure of the persistence and magnitude of their influence. There is no evidence that the effects of the previously considered variables on the rate of imitation are due to the operation of these factors. When these other factors are included, the coefficients of π_{ij} and S_{ij} and the interindustry differences remain relatively unchanged (though the coefficient of S_{ij} becomes nonsignificant).

5. A Stochastic Version of the Model

In this section, we present and test a somewhat more sophisticated stochastic version of the model. For the jth innovation in the ith industry, let $P_{ij}(t,k)$ be the probability that any one of the "hold-outs" at time t will introduce it by time $t + k$, and assume (for small k) that

$$\text{(19)} \qquad P_{ij}(t,k) = \theta_{ij} \frac{m_{ij}(t)}{n_{ij}} k,$$

$$\text{(20)} \qquad \theta_{ij} = b'_i + a'_5\pi_{ij} + a'_6S_{ij} + z''_{ij},$$

where the coefficients in equation (20) are analogous to those in equation (12). To see how closely this resembles the deterministic model, note that the *expected* increase in the number of users between time t and time $t + 1$,

$$[n_{ij} - m_{ij}(t)]\theta_{ij} \frac{m_{ij}(t)}{n_{ij}},$$

is almost identical with the expression given before for the *actual* increase in

[29] We used the National Bureau's reference dates (given in G. Moore, "Measuring Recessions," *Journal of the American Statistical Association* [June 1958]) to determine whether t^*_{ij} was a year of contraction or recovery. The residuals in Figure 7.2 do not seem to be affected by whether or not an innovation was being accepted some time during the depression of the 1930's.

the number of users (see equations (4) and (5)). The only (apparent) difference is that the terms corresponding to Q_{ij} are assumed to be zero from the start here, whereas in section 2 this followed from equation (8).[30]

Whereas our task in the deterministic model was to determine how the actual number of users grows over time, our problem here is to see how the expected number grows. To do so, we first obtain an expression for $P_{ij}^r(t)$—the probability that at time t there are exactly r firms in the ith industry that have not yet introduced the jth innovation. From equation (19),

$$P_{ij}^r(t+k) = P_{ij}^r(t)\left[1 - r\frac{\theta_{ij}}{n_{ij}}(n_{ij}-r)k\right] + P_{ij}^{r+1}(t)(r+1)\frac{\theta_{ij}}{n_{ij}}(n_{ij}-r-1)k + o(k),$$

where $o(k)$ represents terms that, if divided by k, tend to zero as k vanishes.[31] And if we subtract $P_{ij}^r(t)$ from both sides, divide by k, and let k tend to zero, the following differential-difference equation results:

$$\text{(21)}\quad \dot{P}_{ij}^r(t) \begin{cases} = -\dfrac{\theta_{ij}}{n_{ij}} r(n_{ij}-r)P_{ij}^r(t) + \dfrac{\theta_{ij}}{n_{ij}}(r+1)(n_{ij}-r-1)P_{ij}^{r+1}(t), & \text{if } r < n_{ij}-1, \\ = -\dfrac{\theta_{ij}}{n_{ij}} rP_{ij}^r(t), \quad \text{if } r = n_{ij}-1, & \end{cases}$$

with the initial conditions that $P_{ij}^r(t^*_{ij})$ equals unity for $r = n_{ij} - 1$ and zero otherwise, and where $\dot{P}_{ij}^r(t)$ is the time derivative of $P_{ij}^r(t)$, and t^*_{ij} is the date (taken as given)[32] when the innovation was first introduced. From equation (21), it follows [8] that $M_{ij}(t^*_{ij} + V)$—the expected number of firms using the innovation at time $t^*_{ij} + V$—equals

$$\text{(22)}\quad n_{ij} - \sum_{r=1} \frac{(n_{ij}-1)!}{(n_{ij}-r-1)!(r-1)!}\left[(n_{ij}-2r)^2 - \frac{\theta_{ij}}{n_{ij}}V + 2 - (n_{ij}-2r)\sum_{u=r}^{n_{ij}-r-1} u^{-1}\right] e^{-r(n_{ij}-r)(\theta_{ij}/n_{ij})V},$$

[30] To simplify things, we also assume from the beginning that the coefficients of π_{ij} and S_{ij} do not vary among industries (Cf. equation [20]). In the deterministic model, we introduced this assumption later in the analysis.

[31] See note 31 of E. Mansfield, "Technical Change and the Rate of Imitation," *Econometrica* (Oct. 1961), for a derivation of this expression. Note two things: First, the assumption that the probability in equation (19) is the same for all "hold-outs" is obviously only a convenient simplification. Without it, the analysis becomes extremely difficult. Second, we assume that the decisions of the "hold-outs" between time t and time $t + k$ are independent. Hence, the probability that one "hold-out" will introduce it between time t and time $t + k$ is $[n_{ij} - m_{ij}(t)]\theta_{ij}m_{ij}(t)k/n_{ij} + o(k)$, and the probability that none will introduce it is $1 - [n_{ij} - m_{ij}(t)]\theta_{ij}m_{ij}(t)k/n_{ij} + o(k)$.

[32] For all but the tin container, we use December 31 of the year during which the innovation was first introduced by one of these firms (or if we know the month when it was first introduced, we use that December 31 which was closest in time to it). For the tin container, we use February 1, 1935. See Table 7.1 for the year in which each t^*_{ij} falls.

where r runs up to $(n_{ij} - 1)/2$ for n_{ij} odd and up to $n_{ij}/2$ for n_{ij} even.[33]

Thus, the stochastic version of the model leads to the following two propositions: First, the expected number of firms having introduced an innovation at any date subsequent to t^*_{ij} should be given by equation (22). Second, θ_{ij}—the parameter that, for given n_{ij}, determines the expected rate of imitation—should be a linear function of π_{ij} and S_{ij}, the intercept of the function differing among industries.

To see how well the first proposition seems to hold, we estimate θ_{ij}, insert it, n_{ij}, and t^*_{ij} into equation (22), and compare the growth over time in the computed number of users with the actual growth curve. Table 7.1 contains a rough measure of "goodness-of-fit": The root-mean-square deviation of the actual from the computed number of users.[34] Although the agreement between the actual and computed growth curves is sometimes quite good, Table 7.1 shows that equation (22) is never so accurate as equation (9) in representing the data. Perhaps this is due in part to differences in the way parameters are estimated as well as differences in the model.[35]

To test the second proposition, we assume that errors in the estimates of θ_{ij} are uncorrelated with π_{ij} and S_{ij} and hence that

$$\text{(23)} \qquad \hat{\theta}_{ij} = b'_i + a'_5\pi_{ij} + a'_6 S_{ij} + z_{ij}''',$$

where $\hat{\theta}_{ij}$ is our estimate of θ_{ij} and z_{ij}''' is a random error term. Then, proceeding as we did in sections 3 and 4, we get much the same sort of results as those obtained there. (1) We estimate the parameters in equation (23), find that these estimates are almost identical with those given in equation (14) for the analogous parameters,[36] and conclude that the resulting equation represents the data very well ($r = 0.997$). (2) We introduce the additional variables discussed in section 4 into equation (23) and find that their coefficients have the expected signs, but that (except for the coefficient of t_{ij}) all are nonsignificant.

Thus, the model, in either its deterministic or stochastic form, seems to represent the empirical results in Figure 7.1 quite well. Although equation (22)

[33] When n_{ij} is even, some adjustment has to be made to this expression. See N. Bailey, "Some Problems in the Statistical Analysis of Epidemic Data," *Journal of the Royal Statistical Society*, B, 17, 1955, for the adjustment and for a much more extensive discussion of the argument.

[34] To estimate θ_{ij}, we used the statistic suggested by Bailey (in "Some Problems in Analysis of Data," p. 41) to estimate θ_{ij}/n_{ij} and multiplied the result by n_{ij}. This estimate of θ_{ij} is biased, but we can show that the bias is approximately equal to $\theta_{ij}/(n_{ij} - 1)$, which is usually quite small. These estimates, together with the tables of Mansfield and Hensley, were used to calculate the growth over time in the expected number of users (see E. Mansfield and C. Hensley, "The Logistic Process: Epidemic Curve and Applications," *Journal of the Royal Statistical Society*, B, 22, 1960). The root-mean-square errors are computed in the way described in note 16.

[35] Two parameters (l_{ij} and ϕ_{ij}) are fitted to the data in the deterministic model whereas only one (θ_{ij}) is fitted here. However, this is only part of the explanation (see note 37). Note too that the differences between the computed and actual curves are by no means random for a number of these innovations. The computed curves sometimes lie below the actual ones.

[36] The only appreciable difference is that the coefficient of π_{ij} is 0.580 rather than 0.530.

does not fit the data as well as equation (9) in the deterministic version, the purpose and interpretation of these equations are quite different, and hence this may not be very surprising.[37] With regard to the explanation of differences in the rate of imitation, equation (23), like equation (13) in the deterministic model, provides an excellent fit. The stochastic version of the model seems somewhat more reasonable to me, but it is difficult to tell at this point whether it is a significant improvement over the simpler, deterministic version.

Before concluding, one final point should be noted regarding the dispersion about the expected value in equation (22). This dispersion is often relatively large and hence a prediction of the rate of imitation based on equation (22) would be fairly crude, even if the model were correct and θ_{ij} were

Table 7.2. *Expected Value and Standard Deviation of the Number of Firms Having Introduced an Innovation, 1924–1939*

DATE (Dec. 31)	EXPECTED NUMBER	STANDARD DEVIATION
1924	1.0	—
1925	1.5	0.9
1926	2.1	1.4
1927	2.7	1.6
1928	3.6	2.3
1929	4.5	2.7
1930	5.5	2.9
1931	6.4	3.1
1932	7.2	3.2
1933	8.0	3.1
1934	8.8	2.9
1935	9.4	2.6
1936	10.0	2.4
1937	10.5	2.1
1938	10.8	1.8
1939	11.1	1.5

SOURCE: See E. Mansfield and C. Hensley, "The Logistic Process: Epidemic Curve and Applications," *Journal of the Royal Statistical Society*, B, 22, 1960. Small errors due to interpolation are present in both the expected values and standard deviations.

NOTE: Assume that $n_{ij} = 12$, $t^*_{ij} = 1924$, and $\theta_{ij} = 0.42$.

known in advance. To illustrate this, consider an innovation where $n_{ij} = 12$, $t^*_{ij} = 1924$, and $\theta_{ij} = 0.42$. Table 7.2 shows the expected number of firms

[37] In the deterministic model, we fit equation (9) to the data in such a way as to minimize differences between the actual and computed curves. In this model, we use the data to estimate the curve of *averages* that would result if the process were repeated indefinitely. Whereas the point in the former case was to minimize differences between the computed and actual curves, this was not the point here and hence it is not surprising that the deterministic model seems superior in this respect.

having introduced it at various points in time and the relatively large standard deviation of the distribution about this expectation. Of course, despite this variation, the model could set useful upper and lower bounds on how long the entire process would take.[38]

6. Limitations

In evaluating these results, the limitations of the data, methods, and scope of the investigation must be taken into account. For example, although its scope far exceeds that of the few studies previously conducted, it nonetheless is limited to only four industries and to only a few innovations in each one. Before concluding, it seems worthwhile to point out some of these limitations in more detail.

First, there is the matter of scope, Although the four industries included here vary with respect to market structure, size of firm, type of customer, etc., they can hardly be viewed as a cross section of American industry. For example, none is a relatively young industry with a rapidly changing technology. In addition, the innovations included here are all important, they generally required fairly large investments, and the imitation process was not impeded by patents. To check and extend these results, similar studies should be carried out for other industries and other types of innovations. Moreover, international comparisons might be attempted.

Second, the data are not always as precise as one would like. For example, much of the data regarding the profitability of installing an innovation had to be obtained from questionnaires and from interviews with executives of the firms. Although these estimates were checked against others obtained from suppliers of the equipment, trade journals, etc., they are probably fairly rough. Similarly, the estimates of the pay-out period required for investments and the durability of old equipment are only rough. Because of these errors of measurement, the regression coefficients in equations (14) to (18) are probably biased somewhat toward zero.[39]

Third, the data measure the rate of imitation among large firms only. As noted above, firms not exceeding a certain size (specified in Appendix C) were excluded. Moreover, the rate at which firms imitated an innovator, not the rate at which a new technique displaced an old one or the rate at which investment in a new technique mounted, is considered here. Although these related topics are touched on elsewhere in this book, they were not taken up in this chapter.

[38] For further discussion of the distribution about these expected values, see Mansfield and Hensley, "The Logistic Process," and the literature cited therein.

[39] It can easily be shown that measurement errors, if random, have this effect. Another limitation of the analysis is that the model can only be expected to hold for relatively profitable innovations. Certainly, it will not hold in cases where $\pi_{ij} < 1$ and it may do poorly if π_{ij} is not appreciably greater than unity. But this limitation is not very serious because if π_{ij} does not exceed unity, the innovation almost certainly will not become generally accepted, and if π_{ij} is not much greater than unity, the innovation is probably not very important.

Fourth, various factors other than those considered here undoubtedly exerted some influence on the rate of imitation. Variation over time in the profitability of introducing an innovation (due to improvements, the business cycle, etc.) is one potentially important factor that is omitted.[40] Others are the sales and promotional efforts made by the producers of the new equipment and the extent of the risks firms believed they assumed at first by introducing an innovation.[41] I could find no satisfactory way to measure them, but perhaps further research will obviate this difficulty and disclose the effects of these and other such factors on the rate of imitation. I suspect that the results underestimate their influence and that data for more innovations would show that the unexplained variation (due to their effects) is greater than is indicated in Figure 7.2.

7. Summary and Conclusions

The principal conclusions of this chapter are as follows: First, we have constructed a simple model to help explain differences among process innovations in the rate of imitation. This model is built largely around one hypothesis—the probability that a firm will introduce a new technique is an increasing function of the proportion of firms already using it and the profitability of doing so, but a decreasing function of the size of the investment required.[42]

Second, when confronted with data for twelve innovations, this model seems to stand up surprisingly well. As expected, the rate of imitation tended to be faster for innovations that were more profitable and that required relatively small investments. An equation of the form predicted by the model can explain practically all of the variation among the rates of imitation.

[40] One cannot tell with any accuracy how, and to what extent, the profitability varied in each of these cases. But, to illustrate the factors at work, note three broad changes that occurred in the case of three of the innovations. First, the appearance of improved diesel locomotives in the 1930's and 1940's increased the profitability of introducing them and hastened their acceptance. In addition, first costs declined, the relative prices of coal and oil changed, and the composition of the "hold-outs" varied. Second, the demand for toluene during World War I undoubtedly accelerated the imitation process in the case of the by-product coke oven. Third, the Depression reduced the profitability of introducing centralized traffic control, whereas the wartime traffic boom increased its profitability. For further discussion, see note 21. Of course, the relative profitability may have varied less than the absolute profitability.

[41] In each of these cases, there was considerable promotional effort by the producers of the new equipment and considerable help given to the firms that used it. For example, Semet-Solvay and Koppers helped to finance by-product coke plants and provided personnel, and General Motors helped to train diesel operators, kept maintenance men at hand, etc. For some discussion of the perceived risks, see note 7. Another factor that could be very important for some innovations is the presence of patents. Still another is the development of a marked improvement in the innovation. For example, an innovation may be applicable at first to only a few firms and a significant improvement is required before others use it. In a case like this, the model is clearly not applicable. In each of these cases, important improvements occurred, but they were of a more gradual and less significant nature. See Chap. 8. For further discussion of various factors, see E. Mansfield, *The Economics of Technological Change*, Chaps. 4 and 5.

[42] This is the stochastic version of the model. In the deterministic version, we use the proportion of "hold-outs" that introduce it rather than this probability.

Third, there were also interindustry differences in the rate of imitation. We have too few industries to test accurately the hypothesis often advanced that the rate of imitation is faster in more competitive industries, but the differences seem to be generally in that direction.

Fourth, when several other factors are included in the model, the empirical results are largely inconclusive. There was some apparent tendency for the rate of imitation to be higher when the innovation did not replace very durable equipment, when the firms' output was growing rapidly, and when the innovation was introduced in the more recent past. But it was almost always statistically nonsignificant.

CHAPTER 8

The Speed of Response of Individual Firms

The previous chapter, concerned entirely with the over-all rate of imitation, tells us nothing about the factors determining how rapidly a particular firm will respond to a new technique. Can we construct models to help predict whether one firm will be quicker than another? What are the quantitative effects of various factors on a firm's speed of response? Do the same members of an industry tend to be relatively quick to introduce various new techniques, or are the leaders in one case likely to be the followers in another?

The purpose of this chapter is to help answer these questions. First, hypotheses are presented regarding the effects of various factors on the length of time a firm waits before using a particular technique. Second, using data concerning the diffusion of fourteen major innovations, we test these hypotheses and estimate the effects of these factors on a firm's speed of response. Third, using the same data, we estimate the extent to which leadership of this sort in four important industries has been concentrated in the hands of a relatively few firms.

Sections 1, 2, and 3 take up the effects of a firm's size and the profitability of the investment in the innovation on its speed of response. In section 4, we deal with the effects of a firm's growth rate, past profits, liquidity position, and other such factors. Section 5 measures the extent to which leadership of this kind was concentrated among a relatively few firms in the bituminous coal, railroad, brewing, and iron and steel industries.

1. Size of Firm and the Profitability of its Investment in the Innovation

This section presents four propositions regarding the effects of a firm's size and the extent of the returns it can obtain from the innovation on how long it waits before introducing the innovation. If they hold, these propositions should provide a basis for predicting whether one firm will be quicker than another to use an innovation. Section 2 converts these general propositions into testable hypotheses, and section 3 carries out the appropriate tests.

The first proposition states that, other things held equal, the length of time a firm waits before using a new technique tends to be inversely related to its size. Because they have more units of any particular type of equipment, large firms are more likely at any point in time to have some units that will soon have to be replaced. Thus, if an innovation occurs that is designed to replace this type of equipment, they probably can begin using it more quickly than smaller firms. Moreover, large firms, because they encompass a wider range of operating conditions, have a better chance of containing those conditions for which the innovation is applicable at first. This is important because when an innovation first appears its application is sometimes restricted to certain operating conditions, and improvements occur later that extend its usefulness.[1]

For these reasons, the larger firms would be expected to use a new technique more quickly, on the average, than the smaller firms, even if the larger firms were no more progressive than the smaller ones. (A simple proof of this point is given in the following two paragraphs.) This tendency is strengthened or offset by other differences between large and small firms. The larger financial resources, bigger engineering departments, and other advantages of the large firm strengthen it, whereas the unwieldiness and conservatism sometimes associated with increases in firm size tend to offset it. On balance, it seems extremely unlikely that the large firms are so much more sluggish than the smaller ones that this tendency will be offset completely.[2]

The second proposition elaborates on the first. It states that, as a firm's size increases, the length of time it waits tends to decrease at an increasing rate. This proposition stems from the following model. Suppose that a new type of equipment is put on the market and that the jth firm will eventually own α_j units of this equipment. Suppose that x_{ij}, the length of time that elapses (from the date when the innovation is first put on the market) before the jth firm's i^{th} unit ($i = 1, \ldots, \alpha_j$) is installed, is a random variable with cumulative distribution function, $F(x)$, and that the time elapsing before one of its units is installed is independent of that for another unit.[3]

Under these highly simplified circumstances, the expected length of time a firm with an eventual complement of α units will wait before beginning to use the innovation is

$$\text{(1)} \qquad \overline{X}_\alpha = \alpha \int_0^M x[1 - F(x)]^{\alpha-1} F'(x)\, dx,$$

where M is the maximum value of x_{ij}. Integrating by parts, we have

[1] With regard to the innovations used in the empirical work in this chapter, there were important improvements in a number of cases. However, the innovations were applicable to a wide range of firms throughout the period. See note 41, Chap. 7.

[2] For relevant discussions, see C. Carter and B. Williams, *Industry and Technical Progress* (New York: Oxford University Press, 1957); T. Scitovsky, "Economies of Scale and European Integration," *American Economic Review* (Mar. 1956); G. Stocking, *Testimony Before Subcommittee on Study of Monopoly Power*, Judiciary Committee, House of Representatives, 1950; and J. Bohlen and G. Beal, *How Farm People Accept New Ideas*, Special Report No. 15, Agricultural Extension Service, Iowa State College, 1955; as well as E. Mansfield, *The Economics of Technological Change* (New York: Norton, 1968), Chap. 4.

[3] The assumption that $F(x)$ is the same for each unit and that the x's are independent is not very realistic, but small departures from it should make little difference. The results are used only to get a rough idea of the shape of the relationship between a firm's size and its speed of response.

$$\text{(2)} \qquad \overline{X}_\alpha = \int_0^M [1 - F(x)]^\alpha \, dx,$$

from which it is easy to show that

$$\text{(3)} \qquad \overline{X}_\alpha - \overline{X}_{\alpha+1} = \int_0^M [1 - F(x)]^\alpha F(x) \, dx > 0,$$

$$\text{(4)} \qquad (\overline{X}_\alpha - \overline{X}_{\alpha+1}) - (\overline{X}_{\alpha+1} - \overline{X}_{\alpha+2}) = \int_0^M [1 - F(x)]^\alpha F^2(x) \, dx > 0.$$

Thus, if α_j is proportional to the jth firm's size (measured in terms of sales or production), the expected length of time a firm waits should decrease at an increasing rate with increases in its size.

The third proposition states that, other things held equal, the length of time a firm waits tends to be inversely related to the extent of the returns it obtains from the innovation. If these returns are very high, the expected returns are likely to be high enough to make the gamble involved in introducing the innovation seem worthwhile at the outset. If they are not so high, the firm will wait until the risks are reduced to the point where the investment seems warranted.[4]

Finally, the fourth proposition states that as the profitability of a firm's investment in the innovation increases the length of time the firm waits decreases at an increasing rate. Certainly it seems plausible that an increase of one percentage point in the rate of return from the investment in the innovation will have a progressively smaller effect on the firm's rate of response, as the investment becomes more and more profitable.

2. Model and Data

To permit testing, we translate these propositions into the following, more specific model:

$$\text{(5)} \qquad d_{ij} = Q_i H_{ij}^{a_{i2}} S_{ij}^{a_3} e^{\epsilon_{ij}},$$

where d_{ij} is the number of years the jth firm waits before beginning to use the ith innovation, S_{ij} is its size, H_{ij} is a measure of the profitability of its investment in the innovation, ϵ_{ij} is a random error term, and both a_{i2} and a_3 are negative. Q_i is a scale factor that differs from innovation to innovation.

This functional form, although arbitrary in many respects, seems quite reasonable. It is in accord with the propositions in the previous section

[4] See Z. Griliches, "Hybrid Corn: An Exploration in the Economics of Technological Change," *Econometrica* (Oct. 1957); R. Mack, *The Flow of Business Funds and Consumer Purchasing Power*, Columbia, 1941; A. Sutherland, "The Diffusion of an Innovation in Cotton Spinning," *The Journal of Industrial Economics* (Mar. 1959); and Chap. 7 of this volume.

stating that, as S_{ij} and H_{ij} increase, $E(d_{ij})$ decreases at an increasing rate. Moreover, a multiplicative relationship makes sense here since the effect on d_{ij} of each of the exogenous variables is likely to depend on the level of the other. For example, differences in firm size would be expected to have less effect if an innovation is extremely profitable than if it is less so. Similarly, differences in the profitability of the investment in the innovation would be expected to have less effect if a firm is large than if it is small.[5]

To test the propositions in the previous section and to estimate the effects on d_{ij} of S_{ij} and H_{ij}, we estimated a_{i2} and a_3 for a number of innovations, assuming that the latter were in some sense representative. In particular, data were collected regarding the diffusion of fourteen innovations in the bituminous coal, iron and steel, brewing, and railroad industries: the shuttle car, trackless mobile loader, and continuous mining machine (in bituminous coal); the by-product coke oven, continuous wide-strip mill, and continuous annealing (in iron and steel); the pallet-loading machine, tin container, and high-speed bottle filler (in brewing); and the diesel locomotive, centralized traffic control, mikado locomotive, trailing-truck locomotive, and car retarders (in railroads).[6]

Three kinds of data were collected in each case. First, we obtained the date when each major firm in the industry began to use the innovation, and we subtracted it from the date when the first firm began to use it to obtain d_{ij}.[7] Second, an estimate was made of each firm's size. Physical output

[5] Other functional forms were used to see whether the results were sensitive to the particular form used. First, d_{ij} was assumed to be a linear function of P_{ij} (the surrogate described below for H_{ij}) and $\ln S_{ij}$. Second, $\ln d_{ij}$ was assumed to be a linear function of S_{ij} and P_{ij}. In both cases, all the regression coefficients were allowed to vary from one innovation to another. The results were qualitatively similar to those obtained below.

Note that a_3 may differ somewhat from innovation to innovation, depending on the size of the investment required to introduce the innovation and the average profitability of the investment in the innovation. We assume that the differences in a_3 among these innovations are relatively small.

[6] The mikado and trailing-truck locomotive could not be included in Chap. 7 because no data were available regarding π_{ij}. Otherwise, the sample of innovations is the same as in the previous chapter. For descriptions of these innovations, see the references in note 2, Chap. 7, and K. Healy, "Regularization of Capital Investment in Railroads," *Regularization of Business Investment* (National Bureau of Economic Research, 1954).

[7] For the continuous wide-strip mill, all firms having more than 140,000 tons of sheet capacity in 1926 were included; for the by-product coke oven, all firms having more than 200,000 tons of pig-iron capacity in 1900 were included; and for continuous annealing, the nine major producers of tin plate in 1935 were included. For centralized traffic control and car retarders, all Class I railroads with over 5 billion freight ton-miles in 1925 were included. For the mikado and the trailing-truck locomotive, the roads included in Healy's sample ("Capital Investment in Railroads") were included. All Class I railroads in 1925 were included for the diesel locomotive. Firms producing over 4 million tons of coal in 1956 were included for the coal innovations, and firms with more than $1 million in assets in 1934 were included for the brewing innovations. These lower limits on size were imposed because of difficulties in obtaining information concerning smaller firms and because in some cases they could not use the innovation in any event. As it is, data could not be obtained for all of these firms because some went out of business or refused to cooperate. But in most cases the results are complete—or very nearly so. For a more detailed discussion, see Chap. 7. The date when each firm first introduced the innovation was obtained from trade journals,

was used in the coal and brewing industries, ingot capacity was used in steel, and freight ton-miles were used in the railroad industry to measure S_{ij}.[8]

Third, because it was impossible to get a direct estimate of the innovation's profitability to each firm, surrogates were obtained where possible. For example, since the profitability of a firm's investment in a continuous mining machine was likely to vary directly with the percent of its output derived from "high seams," this percentage was used as a surrogate. Other surrogates were the reciprocal of the percent of a railroad's mileage that was double track (centralized traffic control), the reciprocal of the percent of a firm's revenues derived from hauling coal (diesel locomotive), the ratio of a firm's rolling capacity to its ingot capacity (continuous wide-strip mill), and a firm's tinplate capacity as a percent of its ingot capacity (continuous annealing). Despite considerable effort, no suitable surrogates could be found for the remaining nine innovations, and the surrogates used in these five cases are obviously very rough.[9]

---

industry directories, and correspondence with the firms. There were a very few firms for which no data could be obtained and they had to be excluded. The specific sources of these data are listed in Appendix C. Note that these are the dates when firms first introduced the innovation—regardless of the scale on which they did so. The possible objections to this are largely removed by the fact that most of these innovations had to be introduced on a large scale. And in the case of the diesel locomotive, one of the few cases where the innovation did not have to be introduced on such a large scale, we made sure that the dates of first purchase almost always represented dates when a substantial number of diesel locomotives were bought, not dates when a trivial number were acquired. The only alternative would be to use the date when a firm first used the innovation to produce some specified percentage of its output, and in almost every case it would have been extremely difficult, if not impossible, to obtain such data. For a study of one case in which data of (roughly) this type were available, see Chap. 9. For further discussion of these data, see note 11.

[8] The number of firms for each innovation is 71 (diesel locomotive), 12 (continuous mining machine), 12 (continuous wide-strip mill), 9 (continuous annealing), 23 (centralized traffic control), 11 (shuttle car), 11 (trackless mobile loader), 19 (tin container), 18 (pallet-loading machine), 15 (high-speed bottle filler), 12 (by-product coke oven), 28 (trailing-truck locomotive), 22 (car retarders), and 32 (mikado locomotive). For the railroads, the data on freight ton-miles come from the Interstate Commerce Commission's *Statistics of Railways* and pertain to 1925. For the coal firms, the production data come from *Moody's* and relate to 1939 for the trackless mobile loader and the shuttle car. For the continuous mining machine, they come from *Moody's* and the *Keystone Guide* and relate to 1947–1948. For the by-product coke oven, the pig-iron capacities come from the 1901 *Directory of the American Iron and Steel Institute;* for the continuous wide-strip mill, the ingot capacities come from the 1926 *Directory of the American Iron and Steel Institute;* for continuous annealing, the size data are sales volumes taken from *Moody's* for 1935. For high-speed bottle fillers and the pallet-loading machine, production data come from *Modern Brewery Age* and they pertain to 1955; for the tin container, the data are production estimates for 1940 from the *Brewers Journal* (July 15, 1943).

[9] According to interviews with coal executives, a firm with relatively little coal from 4 to 9 feet high would almost surely not find continuous mining machines as profitable as a firm with most of its coal in that range. The greater profitability of the innovation in this range was also pointed out by Bituminous Coal Research, Inc., in a private report. For each firm, the percent of its capacity in this range was computed from the 1949 *Keystone Coal Buyer's Guide*, and this rather crude measure was used as a surrogate for the profitability of the investment. As noted previously, the introduction of diesel locomotives seemed more profitable for firms that hauled little coal. This factor was often cited in the interviews, and its importance has been stressed by J. Yance in his "Technological Change as a Learning

Finally, we assumed that the surrogate, P_{ij}, was proportional to H_{ij} in each case, and it followed that

$$(5') \qquad ln\, d_{ij} = a_{i1} + a_{i2}\, ln\, P_{ij} + a_3\, ln\, S_{ij} + \epsilon_{ij}.$$

(Of course, it is not necessary to assume that the ratio of the surrogate to H_{ij} was the same for each innovation.) Using equation (5′), one can easily obtain least-squares estimates of the a's in those cases in which data on both S_{ij} and P_{ij} are available. Where data are available for S_{ij} only, it is necessary to omit the second term on the right-hand side of equation (5′), the implicit assumption being that $ln\, S_{ij}$ is statistically independent of $ln\, H_{ij}$. Thus, two regressions are run, one including observations where data for both S_{ij} and P_{ij} are available and the other including observations where only data for S_{ij} are available.

3. Empirical Results

Using these rather crude data and techniques, we obtained estimates of a_{i1}, a_3, and (where possible) a_{i2}. The results, based on 127 (data for both S_{ij} and P_{ij}) and 167 (data for S_{ij} only) observations, are shown in Tables 8.1, 8.2, and 8.3 in rows or columns headed by an "a." In examining the results, somewhat more weight must be given to the estimates based on data for both S_{ij} and P_{ij}, since the other estimates are more likely to be biased and they provide evidence regarding only some, not all, of the hypotheses in section 1.

The estimates in Tables 8.1 and 8.2 provide considerable support for these hypotheses. Both in the cases where data for S_{ij} and P_{ij} are available and in the cases where only data on S_{ij} are available, the estimate of a_3 has the proper sign and is statistically significant. Moreover, the two independent estimates of a_3 in Table 8.1 are surprisingly close. Apparently, the elasticity of delay with respect to firm size is about -0.4. That is,

$$\frac{\delta d_{ij}}{\delta S_{ij}} \frac{S_{ij}}{d_{ij}} \doteqdot -0.4.$$

The estimates of a_{i2} are also quite consistent with the hypotheses presented above. In every case, these estimates have the expected sign, although they

---

Process: The Dieselization of the Railroads" (unpublished, 1957). Moreover, according to railroad officials, centralized traffic control was probably less profitable for roads with little double track. The percent of trackage that was double and the percent of revenues derived from coal come from *Moody's*. They pertain to 1926 and 1935. We also noted previously that, according to the interviews, there were considerable economies of scale for the continuous rolling mill and continuous annealing. For given over-all size, it would therefore seem that those firms that specialized most heavily in sheets or tin plate would have found them most profitable. Moreover, these firms had the most to lose by delay. A firm's sheet capacity divided by its ingot capacity was computed from the 1926 *Directory of the American Iron and Steel Institute;* a firm's tin plate capacity as a percent of its ingot capacity was computed from the 1940 *Directory*. Of course, these measures are very crude.

Table 8.1. *Estimates of the Effects of a Firm's Size, Growth Rate, Past Profits, Age of President, Liquidity, and Profit Trend on Its Delay in Introducing a New Technique, Fourteen Innovations, Coal, Steel, Brewing, and Railroad Industries*

PARAMETER[a]	INNOVATIONS WITH DATA ON P_{ij}	INNOVATIONS WITHOUT DATA ON P_{ij}
Size of firm[b]		
a_3	−0.40[c]	−0.32[c]
	(.06)	(.12)
b_3	−.41[c]	−.80[c]
	(.07)	(.28)
c_3	−.82[c]	−.83[c]
	(.26)	(.29)
θ_3	−.42[c]	—
	(.08)	
Growth rate[d]		
b_4	−.11	.69
	(.18)	(.36)
c_4	−.23	.59
	(.29)	(.36)
θ_4	−.08	—
	(.19)	
Past profits[d]		
b_5	1.80	.01
	(2.68)	(4.84)
c_5	6.30	−1.10
	(6.68)	(5.38)
θ_5	.68	—
	(3.07)	
Age of president[d]		
c_6	−1.12	2.44
	(1.50)	(1.51)
Liquidity[d]		
θ_6	.18	—
	(.17)	
Profit trend[d]		
θ_7	.21	—
	(.20)	

SOURCE: See notes 7 to 9.

[a] For definitions of these parameters, see section 2 (for the a's) and section 4 (for the b's, c's, and θ's).

[b] The units in which a firm's size is measured here are billions of freight ton-miles (railroad innovations), thousands of tons of capacity (strip mill and coke oven), dollars of sales (annealing), production in barrels (brewing innovations), production in tons (shuttle car and mobile loader), and millions of tons produced (continuous mining machine).

[c] Statistically significant at 0.05 probability level.

[d] For the units in which these variables are measured, see section 4. (π_{ij} is measured in percentage points, and A_{ij} is measured in years.)

Table 8.2. *Estimates of the Effects of the Profitability of a Firm's Investment in an Innovation on Its Delay in Introducing the Innovation, Five Innovations, Coal, Steel, and Railroad Industries*

INNOVATION	PARAMETER[a]			
	a_{i2}	b_{i2}	c_{i2}	θ_{i2}
Diesel locomotive	−0.03	−0.02	—	0.00
	(.09)	(.11)		(.11)
Continuous-mining machine	−.28	−.67	−1.75	−.31
	(.60)	(1.75)	(2.13)	(1.81)
Continuous wide-strip mill	−.89[b]	−.85[b]	−1.25[b]	—
	(.25)	(.29)	(.38)	
Continuous annealing	−1.53[b]	−1.55[b]	−1.87[b]	−1.29[b]
	(.33)	(.35)	(.42)	(.42)
Centralized traffic control	−.05	−.02	.16	−.23
	(.31)	(.34)	(.35)	(.28)

SOURCE: See notes 7 and 8.

[a] For definition of these parameters, see section 2 (for a_{i2}), and section 4 (for b_{i2}, c_{i2}, and θ_{i2}). Section 2 describes the units in which the surrogates for the profitability of a firm's investment in the innovation are measured.

[b] Statistically significant at 0.05 probability level.

Table 8.3. *Estimates of a_{i1}, b_{i1}, c_{i1}, and θ_{i1}, Fourteen Innovations, Coal, Steel, Brewing, and Railroad Industries*

INNOVATION	a_{i1}	b_{i1}	c_{i1}	θ_{i1}
Data on P_{ij} available:				
Diesel locomotive	2.58	3.17	—	2.79
Continuous mining machine	3.02	4.84	14.82	2.83
Continuous wide-strip mill	7.38	7.70	16.67	—
Continuous annealing	12.77	13.46	26.70	12.46
Centralized traffic control	2.98	3.62	10.04	2.67
Data on P_{ij} unavailable:				
Shuttle car	5.56	8.46	−.11	—
Trackless mobile loader	6.11	9.92	1.43	—
Tin container	6.35	—	—	—
Pallet-loading machine	5.96	9.13	.15	—
High-speed bottle filler	6.17	9.54	—	—
By-product coke oven	4.29	—	—	—
Trailing-truck locomotive	2.34	−.42	−9.56	—
Car retarders	2.66	−.18	−9.30	—
Mikado locomotive	2.76	—	—	—

SOURCE: See notes 7 to 9.

NOTE: For definitions of these parameters, see section 2 (for a_{i1}) and section 4 (for b_{i1}, c_{i1}, and θ_{i1}).

often are statistically nonsignificant.[10] As one would expect, there seems to be considerable variation among innovations in a_{i2}, the estimated elasticity of delay with respect to the profitability of a firm's investment in the innovation ranging from -0.03 (diesel locomotive) to -1.53 (continuous annealing). That is, $\frac{\delta d_{ij}}{\delta H_{ij}} \frac{H_{ij}}{d_{ij}}$ ranges from -0.03 to -1.53.

In addition, there is evidence of another sort that seems to support these findings. When about thirty executives in these industries were interviewed, their impression seemed to be that these hypotheses usually held in their industries.[11] However, there were some interindustry differences in their impression of the effect of a firm's size. In the coal, brewing, and railroad industries, their almost unanimous impression was that the smaller firms were slower than the large ones to install important new techniques, but in the steel industry many believed the opposite to be true.[12]

Of course, all of this measures the effects of S_{ij} and P_{ij} on the average value of d_{ij}. Although these effects may be substantial, the model may nonetheless be of little use for predictive purposes because of large residual

[10] These results indicate that firms for which the potential returns were highest—because of their physical setups, market situations, etc.—tended to be early users of the innovation. Note that the data are such that the line of causation can not be turned about. These results can not possibly be a mere reflection of the fact that early users, by virtue of their quickness, often enjoy a somewhat higher return. For example, the percent of a firm's revenues derived from coal or the percent of its output derived from high seams could hardly be affected by its speed of response to these innovations. Note that when these calculations were carried out some firms had not yet begun using high-speed bottle fillers. We included them by assuming that they would introduce them in 1963. Of course, this makes the results for this innovation rather arbitrary, but it would also have been misleading to exclude them altogether.

[11] For a description of these interviews, see Chap. 7. In passing, the following three points might be noted, since they help to integrate our present data and findings with those presented in Chap. 7. First, it should be noted that our findings in this chapter that the larger firms tend to lead do not contradict our findings there that the rate of imitation tends to be lower in more highly concentrated industries. The results presented here for a particular innovation are based implicitly on a certain industry structure, and how long a firm of given size waits may depend on this structure (and the extent of concentration). Second, two innovations are included here but not in Chap. 7 because of lack of necessary auxiliary data. Third, the firms included here sometimes differ slightly from those included there. For the sake of greater homogeneity we excluded steel firms in dealing with the coal innovations and excluded the switching roads in the case of the car retarders. We also included all Class I roads for the diesel locomotive. Where size data could not be obtained, firms had to be omitted. For further integration of our results with those obtained in Chap. 7, see note 14.

[12] Note two things here. First, in the railroad, brewing, and coal industries, there was almost complete agreement with respect to innovations requiring a fairly large investment. But for techniques that could be installed very cheaply, there was less agreement that the larger firms lead. Of course, this is quite reasonable. Second, the statement that the largest firms do not lead in the steel industry can be found in congressional hearings and popular business literature as well as in these interviews. See *Fortune*, Mar. 1936; Stocking, *Testimony Before H. of R. Subcommittee*, 1950; and G. Stigler, *Testimony Before Subcommittee on Study of Monopoly Power*, Judiciary Committee, House of Representatives, 1950. Moreover, in Chap. 5 it was shown that the largest steel firms accounted for less than their share of the innovations. However, we are concerned here with whether the larger firms were quicker, on the average, not with whether *some* smaller firms led the large ones. We are interested primarily in predicting a firm's speed of response. See section 6, note 27.

variation. What is the probability that a prediction based on these hypotheses would have held in these cases? Suppose that one firm was X times as large as another, the value of P_{ij} being the same for each firm, and that we predicted the larger firm would be quicker than the smaller one to introduce the innovation. What would have been our chances of being correct?

Naturally, the answer depends on how large X was. If $X = 4$, our chances of being correct were about 0.80, whereas if $X = 2$, our chances were only about 0.65. (To obtain these figures, we assume that ϵ_{ij} was normally distributed with variance σ_e^2. If so, our chances of being right were $U[-a_3 \ln X/\sigma_e\sqrt{2}]$, where U is the unit normal cumulative distribution function. Estimates of α_3 and σ_e were inserted in this expression to obtain the figures given above.)

Thus, so long as the difference between firm sizes is quite large, it appears that predictions of this sort would have had a very good chance of being correct. Moreover, although the situation is somewhat more complicated in cases where predictions are based on differences in values of P_{ij} (S_{ij} being held constant), the results suggest that these predictions would also have had a very good chance of being right if the difference in P_{ij} was large. For example, if the investment in a continuous wide-strip mill was four times as profitable for one firm as another, the probability was about 0.95 that the former firm was quicker than the latter to begin using it.[13]

In conclusion, three additional points should be made regarding the empirical results. First, the results are consistent with the hypothesis that a_{i2} depends on the average profitability of the innovation and on the size of the investment required to install it. This would be expected on the basis of the findings of Chapter 7.[14] Second, the results show that a considerable amount of the variation in $\ln d_{ij}$ can be explained by $\ln S_{ij}$ and $\ln P_{ij}$. For those cases where data for both S_{ij} and P_{ij} are available, almost half of the variation can be explained by these variables.[15] Third, if one does not constrain a_3 to be the same for all innovations, the results indicate that (with only one exception) d_{ij} is inversely related to S_{ij} and P_{ij} in the case of each innovation

[13] One can go through the same sort of procedure as that carried out in connection with S_{ij}, but the added difficulty here is that a_{i2} (unlike a_3) varies from innovation to innovation.

[14] It can be shown that the rate of imitation in the case of the ith innovation is a decreasing function of a_{i2}^2, σ_e^2, the variance of $\ln H_{ij}$, and the variance of $\ln S_{ij}$. Thus the fact that a_{i2} seems to be an increasing function of the average profitability of the investment in the innovation and a decreasing function of the size of the required investment would be expected on the basis of our findings in Chap. 7.

[15] However, because the uncontrolled effects of P_{ij} inflate the residuals, the correlation coefficient is only 0.35 in the case where data on only S_{ij} are available. Had it been possible to include smaller firms, the relationship would probably have been stronger. In only two cases was it possible to extend the analysis on an exploratory basis. For the continuous mining machine, data were collected for all firms producing over 100,000 tons in 1948 from the listings of equipment in the Keystone Coal Buyers Guide. As size decreased, a smaller percentage had installed continuous miners as yet, and those that did had installed them later. For centralized traffic control, data for a somewhat larger group of firms were obtained from Healy (see "Regularization of Capital Investment"), and the results were much the same. Of course, the data used here are less reliable than those on which Tables 8.1 to 8.3 are based.

taken separately. However, the relationships are often statistically non-significant.[16]

4. Growth Rate, Profitability, Age of President, Liquidity, and Profit Trend

Besides a firm's size and the extent of the returns it obtains from the innovation, there are many other factors that may influence its speed of response to an innovation. This section presents and tests propositions regarding the effects of five such factors. If one could be reasonably sure that these propositions would hold, they could be useful—in the same way as the propositions in section 1—in predicting whether one firm would be quicker than another to use an innovation.

The first proposition states that, other factors held equal, the length of time a firm waits before introducing a new technique tends to be inversely related to its rate of growth.[17] The second proposition states that, other factors held equal, the length of time a firm waits before introducing a new technique tends to be inversely related to the firm's profitability.[18] Assuming that these propositions (as well as those in section 1) hold, we have

$$\text{(6)} \qquad ln\, d_{ij} = b_{i1} + b_{i2}\, ln\, P_{ij} + b_3\, ln\, S_{ij} + b_4\, ln\, g_{ij} + b_5\, ln\, \pi_{ij} + \epsilon'_{ij},$$

where π_{ij} is the jth firm's profits as a percent of its net worth during a 3-year period soon after the innovation was first put on the market, g_{ij} is the percentage increase in the jth firm's production or capacity (plus 100) during the (approximate) period during which the imitation process was going on, and ϵ'_j is an error term.[19] According to these propositions, b_{i2}, b_3, b_4, and b_5 should be negative.

[16] See note 5. The one exception was the trailing-truck locomotive, where the relationship between S_{ij} and d_{ij} was direct, but not statistically significant.

[17] See Chap. 7, note 26 for further discussion of the effects of this factor.

[18] Ruth Mack, in *The Flow of Business Funds*, p. 289, quotes one machinery manufacturer as saying that the early purchasers tend to be "either the 'wide-awake progressive companies' which were generally . . . in a strong financial position or the 'do or die' group which decided to play a turn of the wheel and sink or swim thereby." According to him, the first group was the more important.

[19] Specifically, g_{ij} is 100 times the ratio of the sales (or freight ton-miles in the case of railroads) in the terminal year to that in the initial year:

Innovation	Terminal Year	Initial Year	Period for π_{ij}
Continuous mining machine	1958	1948	1946–1948
Shuttle car and trackless mobile loader	1948	1938	1938–1940
Continuous annealing	1956	1936	1939–1941
Continuous wide-strip mill	1936	1926	1926–1928
Bottle filler and pallet loader	1958	1950	1950–1952
Railroad innovations	1949	1925	1925–1927

The period to which π_{ij} pertains, is provided in the final column. The data came from *Moody's* and the *Statistics of Railways*. Of course, only firms for which such data could be

To test each of these hypotheses, we estimate the b's, using all the innovations for which the necessary data on g_{ij} and π_{ij} could be obtained from trade sources and *Moody's* for a reasonable number of firms. The resulting least-squares estimates in Tables 8.1, 8.2, and 8.3, based on 115 (data for P_{ij}) and 74 (no data for P_{ij}) observations, provide no strong evidence in support of these hypotheses.[20] Again, two regressions are run, one for observations from which data could be obtained on P_{ij} and one for the rest. Both estimates of b_5 and one estimate of b_4 have the "wrong" sign. None of the estimates is statistically significant.

Of course, if these hypotheses really hold, the inclusion of other independent variables in equation (6) may make it more apparent. As an experiment, we included A_{ij}—the age of the jth firm's president—as an additional independent variable. It is often asserted that younger managements, being less bound by traditional ways, are more likely than older ones to introduce a new technique relatively quickly. Moreover, in agriculture, there is some evidence that this is the case [15, 54]. Assuming that this proposition (as well as the previous ones) holds, we have

$$(7) \qquad \ln d_{ij} = c_{i1} + c_{i2} \ln P_{ij} + c_3 \ln S_{ij} + c_4 \ln g_{ij} + c_5 \ln \pi_{ij} + c_6 \ln A_{ij} + \epsilon''_{ij},$$

where $c_{i2}, \ldots, c_5$ are negative; c_6 is positive; and ϵ''_j is a random error term.

To test this hypothesis, we estimate the c's using those innovations for which data regarding the exogenous variables could be obtained for a reasonable number of firms. Again, two regressions were run, one for observations where P_{ij} could be estimated and one for the rest. The results, based on 50 (data for P_{ij}) and 68 (no data for P_{ij}) observations, are shown in Tables 8.1, 8.2, and 8.3. They provide no real evidence that g_{ij}, π_{ij}, or A_{ij} had a significant influence on d_{ij}; half of the estimates of c_4, c_5, and c_6 have the "wrong" sign and none is statistically significant.[21]

---

obtained were included in the regressions. (Three innovations were excluded altogether because of lack of data.) Of course, if equation (6)—or equations (7) or (8)—is the true model there is a bias in the estimates because of specification error. But it does not appear that this bias is very important. The estimates of a_{i2} and a_3 are not very different from the corresponding estimates of b_{i2} and b_3.

[20] The correlation coefficients were 0.64 (for innovations where data on P_{ij} were available) and 0.50 (for innovations where such data were not available). The number of firms included for each innovation were 65 (diesel locomotive), 8 (continuous mining machine), 10 (continuous wide-strip mill), 9 (continuous annealing), 23 (centralized traffic control), 9 (shuttle car), 9 (trackless mobile loader), 7 (pallet-loading machine), 6 (high-speed bottle filler), 21 (trailing-truck locomotive), and 22 (car retarders). Note that the data are rather crude. The growth data pertain only to very long periods and the rate of growth within these periods was probably far from smooth. The data on π_{ij} pertain only to a 3-year period.

[21] Although the hypothesis that A_{ij} is an important determinant of d_{ij} seems dubious on a number of counts, the results of the agricultural studies indicated that it was worthy of investigation. Data were obtained from *Moody's*, *Standard and Poor's Register of Executives*, *Who's Who*, etc., regarding the age of the president of each firm when the innovation first was used. Of course, there is a problem of timing here. The correlation coefficients

Finally, we take a somewhat less detailed look at the effects of two other factors—a firm's liquidity and its profit trend. With regard to liquidity, one might expect that more liquid firms would be better able to finance the investment in the innovation and that consequently they might be quicker than less liquid firms to use it. With regard to a firm's profit trend, one might suppose that firms with decreasing profits would be stimulated to search more diligently than other firms for new alternatives [101] and that, other things equal, they might tend to be quicker than others to begin using a new technique.

If these propositions (and those in section 1) hold for a given innovation, we have

$$\text{(8)} \qquad \ln d_{ij} = \theta_{i1} + \theta_{i2} \ln P_{ij} + \theta_3 \ln S_{ij} + \theta_4 \ln g_{ij} + \theta_5 \ln \pi_{ij} + \theta_6 \ln L_{ij} + \theta_7 \ln t_{ij} + \epsilon'''_{ij},$$

where L_{ij} is the average value of the jth firm's current ratio (current assets divided by current liabilities) during the 3 years up to and including the year when the innovation was first used in this country; t_{ij} is the slope of the linear regression of the jth firm's profit rate against time (measured in years) during a 6-year period just before the innovation was first used in this country; $\theta_{i2}, \ldots, \theta_6$ are negative; θ_7 is positive; and ϵ'''_{ij} is an error term.

To test these hypotheses, we estimated the θ's, using all innovations for which data regarding all of the exogenous variables could be obtained for a reasonable number of firms. The results, based on a single regression utilizing 101 observations, are shown in Tables 8.1, 8.2, and 8.3. Half of the estimates $\theta_4, \ldots, \theta_7$ have the "wrong" sign, and none is statistically significant.[22]

Finally, the limitations of the results should also be noted. The data we could obtain are limited in quality and scope; and because they are almost impossible to measure, we had to omit many important factors—the amount of research a firm conducted in the relevant area, the preferences of its

---

were 0.70 (for innovations where data on P_{ij} were available) and 0.51 (for the others). The number of firms included for each innovation was 8 (continuous mining machine), 10 (continuous wide-strip mill), 9 (continuous annealing), 23 (centralized traffic control), 9 (shuttle car), 9 (trackless mobile loader), 7 (pallet-loading machine), 21 (trailing-truck locomotive), and 22 (car retarders).

[22] This part of the chapter benefited from discussions with G. von der Linde. The basic data were obtained by K. E. Knight in a term paper. Profits plus bond interest divided by total sales was used as the profit rate for the railroads and profits less preferred dividends divided by net worth were used for the coal and steel firms. Somewhat different measures might have been used instead, but the results would almost certainly have been about the same. Again, the period of time to which the exogenous variables pertain may not be exactly what one would ideally want here. The correlation coefficient was 0.64, the number of firms included for each innovation was 65 (diesel locomotive), 9 (continuous mining machine), 9 (continuous annealing), and 23 (centralized traffic control).

Note that the results regarding the effects of g_{ij}, π_{ij}, A_{ij}, L_{ij}, and t_{ij} turn out to be essentially the same if we do not constrain their "elasticities" to be the same from one innovation to another. Moreover, the results are much the same if we omit each of these variables, each pair of these variables, and so on, from the regressions.

management with respect to risk, the age of its old equipment, and the extent to which manufacturers of the new equipment could exert pressure on it. (Some of these points are detailed in section 6 of the previous chapter.) The results are only a first step toward understanding the factors determining whether one firm will be quicker than another to use an innovation.

5. Concentration of Technical Leadership

To what extent do the firms that are quick—or slow—to introduce one innovation tend to be quick—or slow—to introduce others as well? The answer to this question is important because it shows the extent to which technical leadership is concentrated in the hands of only a few members of an industry. In this section, we see how closely it has been concentrated among firms in the bituminous coal, railroad, brewing, and iron and steel industries.

As a first step, we note that the coefficient of correlation between how rapidly a firm introduces one innovation and how rapidly it introduces another is likely to be inversely related to the time interval separating the date when the one innovation first appeared from the date when the other innovation first appeared. Put differently, as the time interval separating the appearance of two innovations increases, there is likely to be less tendency for the same firms to be relatively quick—or slow—to introduce both. This seems reasonable because, as time goes on, technical leadership, if at all polarized, is likely to pass from one group of firms to another.[23]

Assuming that this hypothesis is correct and that a linear function is satisfactory, we have

$$\rho_{qr} = v + wt_{qr} + Z_{qr}, \tag{9}$$

where ρ_{qr} is the coefficient of correlation between the length of time a firm waits before introducing the qth innovation and the length of time it waits before introducing the rth innovation, t_{qr} is the interval (in years) between the dates when these two innovations were first used commercially, and Z_{qr} is a random error term.[24]

Estimates of v and w for each industry would allow us to estimate the average correlation coefficient, given that t_{qr} is fixed. To obtain estimates of v and w, we estimated ρ_{qr} for each pair of innovations in a given industry (Table 8.4). Then, since the appropriate analysis of covariance provides no

[23] Firms with aggressive managements often lose their taste for pioneering as those managements grow older or as others take their place; and laggard firms sometimes change their ways because of an injection of new blood and capital. With the passage of time, it becomes increasingly likely that those that were particularly receptive to change in a past era have given up this role to others.

[24] It is impossible for the linear function in equation (9) to be applicable throughout the entire range, but it may be a useful local approximation.

Table 8.4. *Values of ρ_{qr} and t_{qr} for Nineteen Pairs of Innovations, Coal, Steel, Brewing, and Railroad Industries*

PAIR OF INNOVATIONS	ρ_{qr}	t_{qr}	NUMBER OF FIRMS
Continuous mining machine: shuttle car	−0.02	10	9
Continuous mining machine: trackless mobile loader	−.17	13	9
Shuttle car: trackless mobile loader	.54	3	11
By-product coke oven: continuous wide-strip mill	.21	30	4
By-product coke oven: continuous annealing	.18	42	5
Continuous wide-strip mill: continuous annealing	.53	12	7
High-speed bottle filler: tin container[a]	.42	16	8
High-speed bottle filler: pallet-loading machine[a]	.24	3	8
Tin container: pallet-loading machine	−.06	13	15
Centralized traffic control: trailing-truck locomotive	.02	2	21
Centralized traffic control: diesel locomotive	.04	2	23
Centralized traffic control: car retarders	.16	2	21
Centralized traffic control: mikado locomotive	.19	25	22
Trailing-truck locomotive: diesel locomotive	.26	0	22
Trailing-truck locomotive: car retarders	.32	0	20
Trailing-truck locomotive: mikado locomotive	.13	23	21
Diesel locomotive: car retarders	.41	0	22
Diesel locomotive: mikado locomotive	.12	23	23
Car retarders: mikado locomotive	−.29	23	21

SOURCE: See section 5.

NOTE: See section 5 for definitions of ρ_{qr} and t_{qr}. The latter is measured in years. The number of firms in the final column differs from that used in previous sections. Only those firms for which we have data regarding both innovations could be included.

[a] Only firms already using high-speed bottle fillers are included here. See note 10.

evidence to the contrary, we assumed that there were no interindustry differences in w; and using least-squares, we found that

$$\text{(10)} \qquad \rho_{qr} = \begin{Bmatrix} 0.40 \\ 0.30 \\ 0.66 \\ 0.28 \end{Bmatrix} - \underset{(0.004)}{0.012}\, t_{qr},$$

where Z_{qr} is omitted and the figures in braces (reading from top to bottom) pertain to brewing, coal, steel, and railroads. (Note that if the rank, not the product-moment correlation coefficient, had been used the results would have been almost exactly the same as those in equation (10).)

This result indicates at least four things. First, given that two innovations occur within a few decades of each other, one can expect some positive correlation between how long a firm waits before introducing one and how long it waits before introducing the other. Thus, if two innovations are reasonably close together in time, there is generally some tendency for the same firms to be relatively quick—or slow—to introduce both.

Second, although there is some such tendency, technical leadership does not seem to be very highly concentrated in most of these industries. Even if two innovations occur simultaneously, the average value of ρ_{qr} in the bituminous coal, brewing, and railroad industries is only about 0.30; and if the innovations occur 5 years apart, it is only about 0.25. To see the implications of this, suppose that firms are ranked by how quickly they used a particular innovation and that one firm is one-quarter of the way down the ranking, whereas a second firm is three-quarters of the way down the ranking. If another innovation occurred 5 years later and $\rho_{qr} = 0.25$, the probability is only about 0.59 that the first firm would introduce it before the second firm.[25]

Third, although leadership seems to be quite widely diffused in most of these industries, there is one exception—iron and steel. If two innovations in steel occur close together in time, there seems to be a fairly high correlation

[25] Let d_{1j} and d_{1k} be the delays for the jth and k[th] firms in introducing the first innovation. Let d_{2j} and d_{2k} be their delays in introducing the second innovation. We suppose that in general for the ith firm,

$$d_{2i} = Q_0 + Q_1 d_{1i} + V_i,$$

where Q_0 and Q_1 are parameters and V_i is a normally distributed random variable with zero expected value and with standard deviation of σ_v. What is the probability that $d_{2j} < d_{2k}$, given that $d_{1j} < d_{1k}$?

It equals

$$\begin{aligned} &Pr\{Q_0 + Q_1 d_{1j} + V_j < Q_0 + Q_1 d_{1k} + V_k\} \\ &= Pr\{Q_1(d_{1j} - d_{1k}) < V_k - V_j\} \\ &= 1 - U[Q_1(d_{1j} - d_{1k})/\sigma_v\sqrt{2}]. \end{aligned}$$

If d_{1i} was normally distributed and d_{1k} was at the 75th percentile and d_{1j} was at the 25th percentile, $d_{1j} - d_{1k} = 1.34\ \sigma_d$, where σ_d is the standard deviation of the d_{1i}. Thus, the probability equals

$$1 - U[-1.34 Q_1 \sigma_d/\sigma_v\sqrt{2}].$$

But it can easily be shown that $Q_1\sigma_d/\sigma_v = \sqrt{r^2/1 - r^2}$. Thus, we have

$$1 - U[-1.34\sqrt{r^2/2(1 - r^2)}]$$

where r is the coefficient of correlation between d_{1i} and d_{2i}. Inserting 0.25 for r, we get the result (0.59) in the text.

Because the linear approximation in equation (9) is unlikely to hold very well for extreme values of t_{qr}, the estimates of v are probably only rough estimates of the values of ρ_{qr} when t_{qr} is extremely small. But for values of t_{qr} of one or more, the results may be quite good. Finally, although a firm's rate of response to a recent innovation is not very useful in most of these industries as a predictor of its rate of response to a current one, it is as good a predictor as a firm's size when P_{ij} is not held constant. (Of course, this is not surprising since, one important reason for ρ_{qr} being greater than zero is the size effect.) But when P_{ij} is held constant, a firm's size seems to be an appreciably better predicter than a firm's rate of response to a previous innovation.

(about 0.60) between how rapidly a firm introduces one and how rapidly it introduces the other. The higher correlation in steel may be due in part to a more unequal distribution among firms of research expenditures than in other industries. Or it may be due to less variation among innovations in capital requirements, in the sorts of firms for which they are most profitable, and so forth.

Fourth, as we expected, the estimate of w is negative and statistically significant. Thus, as the time interval separating two innovations increases, there is less correlation between a firm's speed of response to one and its speed of response to the other. But it is noteworthy that the correlation coefficient decreases very slowly, an increase in the time interval of one year resulting in a decrease of only about 0.01 in the correlation coefficient.

Of course, one important reason why there is a tendency for the same firms to be the leaders is that the large firms tend to be the leaders and the same firms tend to be large (during periods of relevant length). However, this is by no means the whole story. If a firm's residual from the regression of d_{ij} on P_{ij} (when data are available) and $\ln S_{ij}$ is used, rather than d_{ij}, the results are quite similar to those in equation (10).[26] Thus, if one corrects roughly for a firm's size (and, where possible, for the profitability of the innovation), there is still some tendency (1) for the same firms to be the leaders, (2) for this tendency to be rather weak in all industries but steel, and (3) for this tendency to become weaker as the time interval separating the innovations becomes greater.

6. Summary and Conclusions

The principal conclusions of this chapter are as follows: First, the length of time a firm waits before using a new technique tends to be inversely related to its size and the profitability of its investment in the innovation. For the innovations for which we could obtain data, the elasticity of the delay with respect to size of firm is about −0.40 and the elasticity of the delay with respect to the profitability of the investment ranges from −0.03 to −1.53. If the differences in size or the profitability of the investment in the innovation are substantial, these relationships seem to be quite useful for purposes of prediction.

Second, the fact that the large firms tend to be quicker, on the average, than the small ones does not mean that the large firms were more progressive. Even if the large firms were no more progressive, one would expect them to be quicker, on the average, than the small firms. To illustrate this, consider an industry with two firms, one large (80 percent of the market), one small (20 percent of the market). If the large firm does its share of the innovating

[26] The results are based on the regression of d_{ij} on P_{ij} and $\ln S_{ij}$, rather than $\ln d_{ij}$ on $\ln P_{ij}$ and $\ln S_{ij}$ (as in equation (5′)). It seemed unlikely that they would change significantly if $\ln d_{ij}$ and $\ln P_{ij}$ were substituted for d_{ij} and P_{ij}, so we did not rerun the regressions.

(no more, no less), it will be first in 80 percent of the cases—and it will be quicker on the average than the small firm.[27]

Third, when the size of the firm and the profitability of the investment in the innovation are held constant, there is no significant tendency for the length of time a firm waits to be inversely related to its profitability, its growth rate, and its liquidity, or directly related to the age of its president and its profit trend. Although these factors might seem to be important, their effects are often in the "wrong" direction and they are always statistically nonsignificant. Perhaps these variables are less important than other more elusive and essentially noneconomic variables. The personality attributes, interests, training, and other characteristics of top and middle management may play a very important role in determining how quickly a firm introduces an innovation. The presence or absence of a few men in the right places who believe strongly in the value of the new technique may make a critical difference.

Fourth, technical leadership of this kind has not been very highly concentrated in most of the industries for which we have data. Even if one firm was considerably quicker than another to begin using one innovation, the chance that it will also be quicker to introduce another innovation occurring only 5 years later is not much better than 50–50. Apparently, there is no particular group of firms that consistently exercises leadership of this kind and no particular group that consistently brings up the rear. It would be quite misleading to assume, from just three or four innovations, that the firms that are quick to use them are generally the leaders in this sense.

Fifth, the results support the general proposition that the speed at which a firm responds to an investment opportunity is directly related to the profitability of the opportunity. This proposition, which is akin to the psychological laws relating the speed of response to the extent of the stimulus, has played an important role in Chapter 7 and will be useful in Chapter 9 as well.

[27] This point seems to have been misunderstood by W. Adams and J. Dirlam, in their "Big Steel, Invention, and Innovation" (*Quarterly Journal of Economics*, May 1966), p. 168. Incidentally, contrary to their assertion, the model in Chap. 5 is able to explain (at least partly) the observed behavior of the steel industry described in Chap. 5.

For the results to indicate that the large firms were more progressive than the small firms, the data would have to show that the difference in the speed of response between large and small firms is greater than would be expected if a large firm acted as if it were simply the sum of an equivalent number of small, independent firms. Only in the case of the diesel locomotive is the sample of firms large enough to shed much light on this question. In this case, the difference is larger than would be expected on the basis. Whether this is so in the other cases is by no means obvious. In the steel industry, it certainly seems doubtful.

CHAPTER 9

Intrafirm Rates of Diffusion

Chapters 7 and 8 were concerned entirely with the rate of imitation—the rate at which firms begin to use an innovation. To understand how rapidly a new technique displaces an old one, one must consider both the rate of imitation and the intrafirm rate of diffusion—the rate at which a particular firm, once it has begun to use a new technique, proceeds to substitute it for older methods.[1] Together, the rate of imitation and the intrafirm rates of diffusion determine how rapidly productivity rises in response to the new technique.[2]

This chapter explores the determinants of the intrafirm rate of diffusion.

[1] It seems obvious that productivity in an industry can be regarded as a weighted average of the productivity with the old technique and the productivity with the new, the weights reflecting the extent to which the new technique has replaced the old. (Whether one has in mind labor, capital, or total productivity is irrelevant, although it affects the sort of weights one would use.) Thus, if the productivity with the new technique exceeds that with the old, productivity in the industry will rise as the new technique is substituted for the old. The rate at which it rises depends clearly on the rate of diffusion. And if the diffusion process goes on more slowly than it should, productivity will not rise sufficiently rapidly and output will fall below its potential. (Of course if the diffusion process goes on too rapidly, inefficiencies result as well.) For further discussion, see W. Salter, *Productivity and Technical Change* (Cambridge, 1960), and E. Mansfield, *The Economics of Technological Change* (Norton, 1968). Note that the intrafirm rate of diffusion measures how quickly a firm substitutes the new technique for the old *once it has begun to use the technique*. It does not tell us anything about the speed at which it began to use it. See section 2. Note too that some innovations can only be introduced on such a large scale that the intrafirm rate of diffusion is of little relevance. The firm either adopts the innovation or it does not. In addition, we presume here that there is an old technique that the innovation replaces.

When it is assumed that the new technique will completely displace the old, a reasonable but arbitrary measure of the intrafirm rate of diffusion is the time interval separating the date when the innovation accounts for 10 percent of the firm's output from the date when it accounts for 90 percent of the firm's output. (Of course, 10 and 90 are arbitrary; we could just as well use 20 and 80, as in notes 14 and 24 of Chapter 7.) This sort of measure (which is inversely related to the intrafirm rate of diffusion) is used in section 2. If the new technique will eventually displace the old in B percent of the cases, 0.1B and 0.9B can be used instead of 10 and 90.

[2] Granted that one knows the percent of the firms in the industry that have begun to use the innovation at each point in time and the average percent of output produced with the innovation (or some similar measure of the intrafirm rate of diffusion) by these firms at each point in time, one can simply multiply them to get the corresponding measure of the rate of diffusion in the industry (if the firms are roughly of the same size). The rate at which firms begin to use an innovation is studied in Chap. 7, and the factors determining whether one firm will be quicker than another to begin using it is studied in Chap. 8. Thus, the combined results of these previous chapters and the present one should help to explain the rate of diffusion in the entire industry.

Once they become familiar with an innovation, some firms abandon the older technology and replace it very quickly with the new. Others are much slower to make the transition. Once a firm has begun to use a new type of equipment, what determines how rapidly it goes on to substitute it for another type? To help answer this question, we single out one of the most significant innovations that occurred in the interwar period—the diesel locomotive. We construct and test an econometric model to help explain differences among railroads in the rate at which, once they had begun to dieselize, they substituted diesel motive power for steam. Although this model is rough and highly simplified, it seems to stand up quite well; and with appropriate modification, it is likely to prove useful for other innovations too.[3]

In section 1, we provide a brief historical sketch of the nation-wide displacement of the steam locomotive by the diesel. Section 2 describes the differences among railroads in how rapidly, once they had begun to dieselize, they substituted diesel power for steam. In sections 3 and 4, we present and test a simple model to help explain these differences. Section 5 discusses the effects of some additional factors, and section 6 analyzes utilization rather than ownership data.

1. Nation-wide Substitution of Diesel Power for Steam

We begin by considering the over-all process whereby the diesel locomotive displaced the steam locomotive. The story begins in 1924, when the first diesel locomotive was used in this country—and 11 years after the diesel-electric system was first used in Europe.[4] The early diesel locomotives were heavy, slow, and without much power. By 1930, eleven American roads used them at some point on their properties, but they were usually installed where there was a smoke nuisance or a fire hazard.[5]

[3] Studies of the diffusion process are relatively rare for industries other than agriculture. Some attention was devoted to the diesel locomotive by K. Healy; see his "Regularization of Capital Investment in Railroads," *Regularization of Business Investment* (National Bureau of Economic Research, 1954). Moreover, J. Yance did some unpublished work in 1957 on this innovation ("Technological Change as a Learning Process: The Dieselization of Railroads"). But most of their work pertained to the spread of the diesel locomotive among firms, not to the intrafirm rates of diffusion. Thus, the amount of overlap with the present study is relatively small. For further discussion of the diffusion process, see E. Mansfield, *The Economics of Technological Change* (New York: Norton, 1968), Chapters 4 and 5.

[4] During the 1920's, diesel locomotives were employed in a limited way in many countries, their most extensive use being in places lacking coal supplies. The first operational diesel locomotive made in the United States resulted from the joint efforts of Ingersoll-Rand (which built the engine), General Electric (which made the components), and American Locomotive (which made the structure). It was put in demonstrating service on June 9, 1924, in New York. It was used for demonstration purposes for 18 months and then was used by Ingersoll-Rand for experimental and development purposes. This 60-ton unit was equipped with one 300-hp, 200-kw oil engine-generator set. *Railway Review* (May 8, 1926).

[5] When municipal governments put pressure on the railroads to eliminate the smoke nuisance, they sometimes turned to diesel power because it was cheaper than electrification. For a sketch of the early history of the diesel locomotive outside the United States, and for references to more detailed accounts, see J. Jewkes, D. Sawers, and R. Stillerman, *The Sources of Invention* (New York: St. Martin's Press, 1958). For a sketch of its early history

During the 1930's, diesel locomotives became more important in the United States. In 1933, General Motors came out with an improved locomotive that was smaller, faster, and more powerful than previous types, and in 1934 the era of the diesel "streamliners" began.[6] By 1935, 50 percent of the major American railroads had begun to use diesel locomotives (see Table 9.1),

Table 9.1. *Number of Diesel and Steam Locomotives and Number of Major Railroads Using Diesel Locomotives, U.S., 1925–1959*

YEAR (Dec. 31)	DIESEL LOCOMOTIVES		STEAM LOCOMOTIVES		MAJOR USERS[a]	
	NUMBER	PERCENT OF TOTAL	NUMBER	PERCENT OF TOTAL	NUMBER	PERCENT OF TOTAL
1925	1	[b]	67,713	99.4	1	4.2
1927	14	[b]	64,843	99.2	8	33.3
1929	25	[b]	60,572	98.9	9	37.5
1931	80	0.1	57,820	98.6	10	41.7
1933	85	0.2	53,302	98.3	10	41.7
1935	130	0.3	48,477	97.9	12	50.0
1937	293	0.6	46,342	97.4	14	58.3
1939	639	1.4	43,604	96.5	19	79.2
1941	1,517	3.4	41,911	94.4	21	87.5
1943	2,476	5.5	41,983	92.5	23	95.8
1945	4,301	9.3	41,018	88.7	23	95.8
1947	6,495	14.6	36,942	83.3	23	95.8
1949	12,025	27.8	30,344	70.1	23	95.8
1951	19,014	44.8	22,590	53.2	23	95.8
1953	24,209	65.0	12,274	32.9	23	95.8
1955	26,563	79.2	6,266	18.7	24	100.0
1957	29,137	90.0	2,608	8.1	24	100.0
1959	30,097	95.4	871	2.8	24	100.0

SOURCE: Interstate Commerce Commission, *Statistics of Railways*, 1925–1960.

[a] Only railroads with more than 5 billion freight ton-miles in 1925 are included. This group differs slightly from that used in Chapter 7. See Appendix C for the differences. A railroad is counted as a "user" if it owned one or more diesel locomotives.

[b] Less than 1/10 of 1 percent.

these leaders generally being large firms and firms in which the investment in such locomotives was particularly profitable. In particular, they tended to haul little coal. The "coal roads" were reluctant to install diesel locomotives,

---

in this country, see Healy, "Regularization of Investment in Railroads"; Jewkes, Sawers, and Stillerman, *Sources of Invention;* and Yance, "Technological Change as a Learning Process." For a "biography" of one of the earliest diesel locomotives, see *Trains*, Nov. 1956, pp. 26–28.

[6] For a description of the old GM locomotive, see C. Foell and M. Thompson, *The Diesel Electric Locomotive*, Diesel Publications, Inc., 1946.

since this might alienate their important customers and since coal was relatively cheap for them.[7]

By 1940, according to a small-scale survey conducted in connection with this chapter,[8] most major American railroads seemed to regard the diesel switcher as being completely out of the experimental stages, although there was still considerable uncertainty regarding its maintenance costs and other factors governing its profitability. When the United States entered World War II, the diesel locomotive had gained considerable acceptance for switching and limited acceptance for other purposes. By that time it accounted for about three-fourths of new orders [37].[9]

Defense needs and priorities governed the production and allocation of diesel locomotives during the war. The allocation of diesel power among the nation's railroads was controlled by the Office of Defense Transportation, and the locomotive builders were allocated material by the War Production Board. Because of the wartime increases in traffic and the change in traffic flows, there was a considerable need for new motive power to replace many of the old steam locomotives. At first, materials (particularly for the diesel power plant) were very tight; but as time went on and the need became more pressing, controls were relaxed and the production of diesel locomotives was stepped up considerably.[10] During the war, about 2,800 diesel locomotives were acquired by American railroads, and by the end of 1945, they constituted almost 10 percent of the total locomotive stock.

When the war ended the acceptance of the diesel locomotive was widespread, but few firms expected it to displace the steam locomotive for all types of work. For example, only four of the seventeen firms for which we

[7] There is a definite relationship between how rapidly a firm began using diesels, and the size of the firm and the percent of its revenue obtained from hauling coal. When only these two independent variables are used, the coefficient of correlation for all Class I railroads is 0.70. See Chap. 8.

[8] Letters were sent to the presidents of most of the railroads included in the sample in Table 9.3. These letters asked various questions regarding the firms' acceptance of the diesel locomotive. About two-thirds of the railroad presidents replied.

One question was: when were diesel locomotives no longer considered experimental by your firm. A percentage breakdown of the replies shows: before 1932, none; 1932–1935, 23 percent; 1936–1939, 23 percent; 1940–1943, 39 percent; and 1944–1945, 15 percent. This pertains only to switchers. Road locomotives generally remained in the experimental stage several years longer. Of course, some of the variation here may be due to differences in the interpretation of "experimental." For a summary of some of the other data obtained in this way, see notes 9 and 11.

[9] According to the correspondence cited in note 8, most diesel locomotives introduced before World War II were used for switching. Only about 15 percent of the responding firms used them for any work other than switching at that time. In the course of other correspondence with the firms listed in Table 9.3, each firm was asked to estimate the average pay-out period for diesel locomotives introduced prior to World War II. About half of the firms replied. The unweighted average of the estimates obtained was 6 years. Of course, there was substantial variation of the estimates by individual firms about this average. For comments on the data, see Chap. 7 and Appendix C.

[10] According to the president of a major Eastern road it became much easier for the railroads to obtain diesel locomotives after May 1943.

have data planned at that time to dieselize completely.[11] However, as time went on, several developments helped to make the advantages of complete dieselization more obvious. First, further refinements were made in diesel design, and the price per horsepower of the diesel locomotive continued to decline relative to steam.[12] Second, it became obvious that large savings could be effected by completely eliminating the facilities needed to service and repair steam locomotives. Third, the remaining uncertainties regarding the diesel locomotive's performance and maintenance were largely dispelled, and the problems in training crews and ancillary personnel were met—with the assistance of the locomotive manufacturers.[13]

Between 1946 and 1955, most of the firms for which we have data decided to dieselize completely. Because they were closer to complete dieselization when the decision was made, they generally planned to accomplish it in only 3 or 4 years—rather than in the 10-year period planned by those that made the same decision in 1945. As a rule, these plans were carried out on time, and by 1959 the diesel locomotive had almost completely displaced the steam locomotive.[14]

2. Intrafirm Rates of Diffusion

Although the previous section provides the necessary background, it tells us very little about the rate at which particular railroads, once they had begun to dieselize, substituted diesel motive power for steam. Table 9.2

[11] For eleven of the seventeen firms, we received complete data on Y_1, the year when plans for complete dieselization were first drawn up, Y_2, the year when these plans visualized that complete dieselization would be accomplished, and Y_3, the year when in fact it was accomplished. These data follow:

Railroad	Y_1	Y_2	Y_3	Railroad	Y_1	Y_2	Y_3
1	1951	1954	1954	7	1945	1955	1956
2	1950	1953	1953	8	1954	1958	1957
3	1945	1955	1955	9	1947	1958	1955
4	1945	1955	1953	10	1946	1956	1956
5	1945	1955	1953	11	1947	1961	1958
6	1954	1959	1960				

[12] In the 1920's and early 1930's, the smallest diesel switchers cost about $100,000. By 1936, a 1,000-hp diesel switcher cost about $100,000 and a large steam switcher cost about $70,000. By 1948, a 1,000-hp diesel switcher still cost about $100,000 but a large steam switcher cost about $120,000. See Healy, "Regularization of Investment in Railroads," p. 175.

[13] Diesel manufacturers helped firms choose the most profitable installations, determine rates of return, and set up schedules. They helped train operating and maintenance personnel. One manufacturer conducted a service by which firms submitted cost data on a confidential basis and received summaries of corresponding data from other firms in return. See Yance, "Technological Change."

[14] A comparison of Y_2 with Y_3 (in note 11) suggests that these programs were generally completed on time.

Table 9.2. *Time Interval Between Date When Diesel Locomotives Were 10 Percent of All Locomotives and Date When They Were 90 Percent of All Locomotives, Thirty Randomly Chosen Class I Railroads*

TIME INTERVAL (years)	NUMBER OF FIRMS[a]	PERCENTAGE OF FIRMS
14 or more	3	10
11 to 13	7	23
8 to 10	11	37
5 to 7	3	10
3 to 4	6	20
Total	30	100

SOURCE: Interstate Commerce Commission, *Statistics of Railways*, 1925–1961.

[a] The railroads included here are listed in Table 9.3. See Mansfield [85].

provides some information on this score. It shows how long it took thirty randomly chosen firms to increase their stock of diesel locomotives from 10 percent to 90 percent of their total locomotive stock.[15] Despite its limitations, this measure is a reasonable first approximation that should do for present purposes, although it will be replaced by a better measure in the following section. In interpreting this measure, note that it does not indicate how rapidly a firm accepted the diesel locomotive, as measured by the rate at which diesels came to dominate the firm's locomotive purchases. Although it is influenced by this factor, it also reflects the rate at which steam locomotives were scrapped and the rate at which locomotives of all kinds were purchased. Since we are interested in the effect of the rate of diffusion on productivity, the latter factors must also be included.[16]

Note too that this measure provides no information regarding the date when various railroads "began" to dieselize. Since Chapters 7 and 8 investigated the factors determining how rapidly a firm begins to use an innovation (the diesel locomotive and others), we simply take these dates as given and

[15] Throughout this chapter, the year when a firm "began" to dieselize will mean the year when diesel locomotives first reached 10 percent of the firm's total locomotive stock. When the quotation marks are omitted, it will mean the year when the first diesel locomotive was purchased.

[16] Three further points might be noted: First, for discussions showing the relevance of the sort of measure used in Table 9.2 (rather than the rate at which diesels came to dominate locomotive purchases) to changes over time in productivity, see note 1 and Salter, *Productivity and Technical Change*. Of course we concentrate entirely on differences in productivity between diesel and steam locomotives and ignore the variation in productivity among locomotives of each type. Second, little or nothing can be deduced from Table 9.2 regarding the rate at which diesel locomotives came to dominate new orders. Third, another measure of the intrafirm rate of diffusion, V_i, is used in the following section. This measure has the advantage that it does not rely on any arbitrary percentages like 10 and 90. For any P_1 and P_2, if the model holds, V_i is inversely proportional to the time interval separating the date when diesels were P_1 percent of the total from the date when they were P_2 percent of the total. Results based on V_i also indicate that there were large differences in the intrafirm rates of diffusion.

focus our attention on how rapidly a firm went on to substitute diesel power for steam. It might be noted, however, that these firms generally reached the point where diesels were 10 percent of their locomotive stock after World War II. Thus, when we deal with the results in Table 9.2, we are dealing almost exclusively with postwar developments.[17]

Finally, returning to Table 9.2, we find from the results that there were substantial differences among firms in the rate of diffusion. Although 9 years were required on the average to increase a firm's stock of diesels from 10 to 90 percent of the total, some firms took 3 years and others took 16. Once they had "begun" to dieselize, why did some firms make the transition so much more quickly than others? The material presented in the previous section is of limited use in answering this question, since it pertains mostly to industry-wide developments and to the period before the end of the war. The following sections present and test a simple model designed to help answer it.

3. A Simple Model

Let $D_i(t)$ be the number of diesel locomotives owned by the ith firm at time t, N_i be the number of steam locomotives owned by the firm before it began to dieselize, and R_i be the number of steam locomotives replaced by a single diesel. If the ith firm's traffic volume and R_i remain approximately constant during the relevant period,[18] the total number of locomotives owned by the ith firm at time t is

$$T_i(t) = N_i - (R_i - 1)D_i(t). \tag{1}$$

Since the firm will therefore employ N_i/R_i diesel locomotives when fully dieselized, there are $[N_i/R_i - D_i(t)]$ places left to be filled with diesels at time t.

Let π_i be the rate of return that the ith firm could obtain by filling one of these places with a diesel locomotive (assume for simplicity that this rate of return is the same for all places and all t), $U_i(t)$ be a measure of the apparent riskiness at time t involved in its making such an investment, S_i be a measure of its size, and C_i be a measure of its liquidity at the time when it "began" to dieselize. Letting

[17] If we are interested primarily in how rapidly a firm went from 10 to 90 percent of full dieselization, it is only important that our model hold for the postwar period. If it does not hold so well for earlier times, the result will only be that it will not explain the movement up to about 10 percent very well.

[18] The assumption that each firm's traffic volume (and motive power requirements) remained constant over time is only a convenient approximation, but during the late 1940's and early 1950's (when dieselization was going on at a significant pace) there was little or no trend in railroad traffic. If 1929 = 100, the index of railroad output was 144 in 1946 and 136 in 1956. On many railroads, there may have been a tendency for R_i to decrease with time. But for most of the relevant period, the decrease was probably slight.

(2) $$W_i(t) = [D_i(t + 1) - D_i(t)]/[N_i/R_i - D_i(t)],$$

we suppose that

(3) $$W_i(t) = f(\pi_i, U_i(t), S_i, C_i, \ldots).$$

The rationale for this hypothesis is as follows. Other things being equal, how heavily a firm invested in diesel locomotives between time t and time $t + 1$ certainly depended on how profitable and how much of a risk such an investment seemed at time t. Thus, one would expect $W_i(t)$—the proportion of unfilled places that were filled with a diesel locomotive during this period—to be directly related to π_i and inversely related to $U_i(t)$.

Because more liquid firms were better able to finance the necessary investment and to take the risks, one might expect them, all other things being equal, to have invested more heavily than other firms. Moreover, smaller firms might have been expected to convert to diesels more rapidly than larger ones because of the costliness of operating two kinds of motive power in a small system, because of the smaller investment (in absolute terms) required to convert, and perhaps because of the quicker process of decision-making in smaller units. Thus, $W_i(t)$ may be directly related to C_i and inversely related to S_i.[19]

Since $U_i(t)$ cannot be measured directly, we assume that

(4) $$U_i(t) = g(L_i, R_iD_i(t)/N_i, \ldots),$$

where L_i is the time interval separating the year when the first firm (in this country) "began" using diesel locomotives from the year when the ith firm "began" using them, and $R_iD_i(t)/N_i$ is the proportion of places in the ith firm already filled at time t. Equation (4) assumes that the longer a firm waited before "beginning" to use the diesel locomotive, the more knowledge it had derived from other firms' experiences with the diesel, and the less uncertainty it had regarding the diesel locomotive's profitability when it "began" to dieselize. It also assumes that the nearer a firm was to full dieselization at time t (that is, the greater was $R_iD_i(t)/N_i$), the less was its uncertainty at time t relative to its uncertainty when it "began" to dieselize.[20] Inserting equation (4) into equation (3), we have

[19] One might also reason that because larger firms were better able to finance the investment and take risks, the effects of S_i might be in the opposite direction from that supposed in the text. Originally I thought that this might be the case but further research has convinced me otherwise. Apparently, these effects are more than counterbalanced by those discussed above. Of course, a larger investment in diesel locomotives between time t and time $t + 1$ does not necessarily result in a higher value of $W_i(t)$. One can only be sure that this will be the case if N_i/R_i and $R_iD_i(t)/N_i$ are held constant, since under these circumstances the number of places left to be filled with diesel locomotives at time t will be constant. Thus, N_i/R_i and $R_iD_i(t)/N_i$ may be two of the factors influencing $W_i(t)$ that are omitted from equation (3). In the analysis below, both are introduced explicitly (N_i/R_i being closely related to A_i).

[20] The reasoning here is as follows. $U_i(t)$ is identically equal to the product of $U_i(t^*_i)$ and $U_i(t)/U_i(t^*_i)$, where t^*_i is the year when the ith firm "began" to use diesel locomotives. We

(5) $$W_i(t) = h(\pi_i, L_i, R_iD_i(t)/N_i, S_i, C_i, \ldots),$$

where, according to the model, $W_i(t)$ is directly related to each of the independent variables other than S_i.[21]

We assume that $W_i(t)$ can be approximated within the relevant range by a quadratic function of $\pi_i, L_i, \ldots, C_i$, but that the coefficient of $[R_iD_i(t)/N_i]^2$ is zero. Then substituting the corresponding differential equation for the difference equation that results and recognizing that $\lim_{t \to -\infty} D_i(t) = 0$, we have

(6) $$D_i(t) = N_i\{R_i[1 + e^{-(\alpha_i + V_it)}]\}^{-1},$$

where

(7) $$V_i = c_1 + c_2\pi_i + c_3L_i + c_4S_i + c_5C_i + \epsilon_i,$$

ϵ_i is a random error term, and c_2, c_3, $-c_4$, and c_5 should be positive if the model holds. The argument leading up to equations (6) and (7) is exactly like that in Chapter 7.[22]

Finally, if $P_i(t)$ is the proportion of the ith firm's locomotives that are diesels at time t,

(8) $$P_i(t) = D_i(t)/T_i(t),$$

and inserting equations (1) and (6) into equation (8), we have

(9) $$P_i(t) = [1 + e^{-(\alpha'_i + V_it)}]^{-1}.$$

---

assume that $U_i(t^*_i)$ is inversely related to L_i and that $U_i(t)/U_i(t^*_i)$ is inversely related to $R_iD_i(t)/N_i$. Thus, $U_i(t)$ is inversely related to L_i and $R_iD_i(t)/N_i$. These assumptions seem reasonable (and the data below bear them out), but their roughness should be obvious, and somewhat different measures might have been used. For example, we might have used $D_i(t)$ rather than $R_iD_i(t)/N_i$, but the latter seems to work quite well.

[21] Equation (5) is quite consistent with interview data and previous studies of the diffusion process. A dozen railroad officials, when interviewed in connection with this study, stressed the importance of each of these independent variables. Descriptions of these interviews are contained in the previous two chapters and Appendix C.

J. Coleman, E. Katz, and H. Menzel present evidence regarding the diffusion of an antibiotic that tends to support the hypothesis that $W_i(t)$ is directly related to L_i ("The Diffusion of an Innovation Among Physicians," *Sociometry* [Dec. 1957]). Z. Griliches presents evidence regarding hybrid corn that tends to support the hypothesis that $W_i(t)$ is directly related to π_i ("Hybrid Corn: An Exploration in the Economics of Technological Change," *Econometrica*, Oct. 1957). J. Yance, in his unpublished "Technological Change as a Learning Process," presents evidence that tends to support the hypothesis that $W_i(t)$ is directly related to $R_iD_i(t)/N_i$. Chaps. 7 and 8 present evidence regarding a dozen innovations that bears on several of these hypotheses.

With regard to the effects of C_i, note too that about half of the respondents to the letters described in note 8 emphasized that the financial condition of the firm was an important determinant of the rate of diffusion of the diesel locomotive. Unfortunately, it is difficult to know precisely what they meant by "financial condition" and how this can be measured. However, from the remarks, it seemed that the liquidity of the firm was closely linked with their notion of "financial condition."

[22] See Chap. 7, section 2.

Thus, if the model holds, the proportion of a firm's locomotives that were diesels should be a logistic function of time. And the parameter of this function measuring the intrafirm rate of diffusion, V_i, should be linearly related to π_i, L_i, S_i, and C_i. In the following section, we see how well this model can explain the observed differences in intrafirm rates of diffusion.[23]

4. Tests of the Model

To test this model, we obtained data regarding $P_i(t)$ from *Statistics of Railways* [59] for the thirty railroads in Table 9.2 for each year from 1925 to 1960.[24] Using ICC data and *Moody's*, we also obtained measures of L_i (the year when the ith firm "began" to dieselize less 1941), S_i (the ith firm's freight ton-miles in 1949), and C_i (the ith firm's average ratio of current assets to current liabilities in the 2 years prior to and including the year when it "began" to use diesel locomotives). Rough estimates of π_i were obtained mainly from correspondence with the firms.[25] The estimates of L_i, S_i, and C_i appear in Table 9.3. The estimates of π_i, which were obtained from the firms with the assurance that their replies would remain confidential, are omitted.

First, we use these data to see how well $P_i(t)$ conforms to a logistic function. If equation (9) holds,

[23] Of course, it would have been preferable to have studied the growth over time in the percent of total traffic hauled by diesel locomotives rather than the percent of the locomotive stock that are diesels. I used the latter measure because it was the only one that was published on a firm-by-firm basis. Section 6 analyzes what few data are available regarding the growth over time in the percent of freight "output produced" by diesel locomotives. Note too that, if the model holds, there is a simple relationship between V_i and the figure for the ith firm in Table 9.2. The latter equals $4.39/V_i$.

[24] An entire system is regarded as a single railroad here.

[25] Each firm in Table 9.3 was asked to estimate the average pay-out period for the diesel locomotives it bought during 1946–1957. The reciprocal of this pay-out period (which is a crude estimate of the rate of return) is used as an estimate of π_i for the twenty-two firms that replied. For the others, we estimated π_i in the following way: First, we estimated R_i (except for a multiplicative constant) for all thirty firms. To obtain this estimate, we assumed that all railroads were working at 100 ϕ percent of "capacity" in 1939 and at 100 θ percent of "capacity" in 1957. Thus, if the ith firm hauled t_i^o ton-miles of freight in 1939 and had Z_i^o locomotives in 1939 (practically all steam), $\phi Z_i^o/t_i^o$ steam locomotives were required per ton-mile of freight on this road. Similarly, if the ith firm hauled t_i^1 ton-miles of freight in 1957 and had Z_i^1 locomotives in 1957 (practically all diesels), $\theta Z_i^1/t_i^1$ diesel locomotives were required per ton-mile of freight on this road. Thus, one diesel locomotive can replace $\phi Z_i^o t_i^1/\theta Z_i^1 t_i^o$ steam locomotives. As a measure of R_i, we use $Z_i^o t_i^1/Z_i^1 t_i^o$—which was proportional to R_i if these assumptions hold. Of course this is very crude. Freight ton-miles are not a completely adequate measure of a firm's output. All firms may not have been operating at approximately the same percentage of "capacity." A firm's "capacity" may not be a linear, homogeneous function of the number of locomotives it owned. The resulting measures of R_i are only rough approximations. Second, we found that our estimate of π_i was correlated with our estimate of R_i for the twenty-two firms where data on π_i were available. According to the interviews, such a relationship would be expected. Using the regression of π_i on R_i, we estimated the value of π_i for the remaining eight firms on the basis of their value of R_i. Note that we assume that the interfirm differences in the profitability of introducing diesel locomotives were approximately the same over time; variation over time during the relevant period is ignored. There is some evidence that the diesel locomotives introduced after the war were somewhat less profitable than those introduced before and during the war (although the very last stages of dieselization were often the most profitable).

Table 9.3. *Estimates of L_i, S_i, C_i, M_i, Q_i, K_i, and A_i*

RAILROAD	L_i	S_i	C_i	M_i	Q_i	K_i	A_i
Pennsylvania	7	54.3	1.68	94	243	4.1	2300
New York Central	7	38.9	1.55	90	237	3.4	1720
Baltimore and Ohio	7	25.3	1.61	95	221	4.5	990
Illinois Central	9	18.7	1.68	92	275	8.3	620
Burlington	3	18.5	1.29	95	309	8.7	690
Missouri Pacific	6	21.2	2.55	97	295	4.9	910
Great Northern	3	16.3	1.44	93	287	6.5	580
Rock Island	0	12.8	2.35	88	312	3.3	400
Northern Pacific	4	11.6	1.98	89	394	4.8	510
Lehigh Valley	2	4.3	1.45	90	191	6.0	180
Nickel Plate	6	9.4	1.73	76	243	5.5	340
Lackawanna	4	4.1	2.17	80	162	2.3	180
Boston and Maine	3	3.1	1.44	96	157	5.2	220
Chicago and Eastern Illinois	5	1.6	1.93	100	193	3.1	60
Duluth, Missabe, and Iron Range	12	3.2	.88	81	76	27.7	130
Denver and Rio Grande	1	5.0	1.31	83	284	5.1	210
Bessemer and Lake Erie	9	2.1	1.98	51	108	6.8	80
Western Pacific	1	3.2	2.37	79	458	4.7	140
Monon	4	1.0	5.52	100	166	5.0	40
Florida East Coast	4	0.8	2.42	100	226	5.3	80
Maine Central	5	0.9	1.46	99	120	4.1	60
Pittsburgh and West Virginia	6	0.4	2.21	62	62	2.1	20
Kansas, Oklahoma, and Gulf	8	0.5	1.71	90	126	5.0	10
Seaboard Air Line	2	7.9	2.61	86	217	10.2	40
Virginian	13	3.2	1.92	50	248	7.9	40
Chesapeake and Ohio	8	27.0	1.26	76	276	5.7	780
Chicago and North Western	4	12.4	1.54	99	180	6.9	510
Norfolk and Western	16	15.3	1.96	65	275	7.2	460
Missouri-Kansas-Texas	6	5.1	1.35	100	303	3.1	190
Union Pacific	4	29.0	2.28	84	559	7.9	1040

SOURCE: Interstate Commerce Commission, *Statistics of Railways*, and *Moody's Railroads.*

NOTES

L_i—To measure L_i, we obtained from the *Statistics of Railways*, the year when each firm's diesel locomotives first reached 10 percent of its total locomotive stock, and we deducted 1941 (the year when diesel locomotives reached 10 percent of the total locomotive stock on the first American railroad).

S_i—The number of freight ton-miles (in billions) in 1949 (obtained from the *Statistics of Railways*).

C_i—The average of the current ratio in the year prior to and including the year when diesel locomotives first reached 10 percent of the firm's total locomotive stock.

M_i—The percentage of a firm's steam locomotives that were 15 years old or more at the time when diesel locomotives first reached 10 percent of the firm's total locomotive stock.

Q_i, K_i, A_i—See section 5 for a definition of Q_i, K_i, and A_i.

Table 9.4. *Estimates of α'_i, $\hat{V}_i$, Coefficient of Determination, and $\hat{H}_i$*

RAILROAD	$\hat{\alpha}'_i$	$\hat{V}_i$	COEFFICIENT OF DETERMINATION[a]	$\hat{H}_i$[b]
Pennsylvania	−7.48	0.43	0.92	0.73
New York Central	−5.95	.35	.91	.88
Baltimore and Ohio	−6.10	.34	.98	—
Illinois Central	−6.21	.30	.92	—
Burlington	−4.80	.29	.99	.49
Missouri Pacific	−6.94	.44	.99	.73
Great Northern	−4.44	.27	.95	.49
Rock Island	−4.32	.29	.87	.63
Northern Pacific	−5.20	.27	.89	—
Lehigh Valley	−4.72	.33	.79	.88
Nickel Plate	−6.47	.34	.93	—
Lackawanna	−4.20	.28	.90	.73
Boston and Maine	−5.01	.33	.94	.63
Chicago and Eastern Illinois	−5.93	.40	.73	2.20
Duluth, Missabe, and Iron Range	−9.39	.40	.71	—
Denver and Rio Grande	−3.96	.25	.93	.44
Bessemer and Lake Erie	−7.18	.41	.74	—
Western Pacific	−4.79	.36	.87	.63
Monon	−11.97	1.04	.84	2.93
Florida East Coast	−5.52	.35	.91	—
Maine Central	−6.91	.45	.95	1.10
Pittsburgh and West Virginia	−8.50	.51	.85	1.10
Kansas, Oklahoma, and Gulf	−11.74	.73	.45	1.46
Seaboard Air Line	−5.28	.36	.97	.55
Virginian	−15.32	.75	.57	—
Chesapeake and Ohio	−10.30	.59	.88	.88
Chicago and North Western	−5.80	.33	.99	.88
Norfolk and Western	−30.48	1.35	.98	—
Missouri-Kansas-Texas	−11.16	.73	.94	1.25
Union Pacific	−5.58	.33	.97	—

SOURCE: Interstate Commerce Commission, *Statistics of Railways*, 1925–1956 and additional ICC data described in note 32.

[a] The square of the coefficient of correlation between $ln\,[P_i(t)/1 - P_i(t)]$ and t, the observations being weighted as Berkson has suggested. See his "A Statistically Precise and Relatively Simple Method of Estimating the Bio-Assay with Quantal Response, Based on the Logistic Function," *Journal of the American Statistical Association*, Sept. 1953. As Z. Griliches pointed out, high correlation coefficients of this sort should be taken with a grain of salt. See his "Hybrid Corn: An Exploration in the Economics of Technological Change," *Econometrica*, Oct. 1957. Nonetheless, the fits seem on inspection to be reasonably good in almost all cases.

[b] See section 6 (and note 32 in particular) for a definition of $\hat{H}_i$ and a description of how it was obtained.

(10) $$ln\,[P_i(t)/1 - P_i(t)] = \alpha'_i + V_i t.$$

Thus, one crude way to measure the goodness of fit of the logistic function is to see how well $ln\,[P_i(t)/1 - P_i(t)]$ can be represented by a linear function of t. Table 9.4 shows that the correlation between these two variables is generally very high. Omitting two cases, the average coefficient of determination (r^2) is 0.90. Thus, the results suggest that a logistic function can represent the data reasonably well.

Second, we test whether V_i conforms to equation (7). Using equation (10), we obtained least-squares estimates of V_i (after weighting the observations appropriately).[26] Assuming that the errors in these estimates are uncorrelated with π_i, L_i, S_i, and C_i, we have

(11) $$\hat{V}_i = c_1 + c_2\pi_i + c_3L_i + c_4S_i + c_5C_i + \epsilon'_i,$$

where $\hat{V}_i$ is the estimate of V_i and ϵ'_i is an error term. Using least-squares to estimate the c's, inserting these estimates into equation (8), and suppressing ϵ'_i, we have

(12) $$\hat{V}_i = -0.163 + \underset{(0.492)}{0.900\,\pi_i} + \underset{(0.008)}{0.048\,L_i} - \underset{(0.0023)}{0.0028\,S_i} + \underset{(0.040)}{0.115\,C_i},$$

where the quantities in parentheses are standard errors.

The results are quite encouraging. The estimates of c_2, c_3, c_4, and c_5 turn out to have the expected signs, and all but c_4 are statistically significant (0.05 level). About 70 percent of the observed variation in $\hat{V}_i$ can be explained by the regression, the correlation coefficient being 0.83. Thus, the model, simple and incomplete though it is, can explain a substantial portion of the interfirm variation in the intrafirm rates of diffusion.

A convenient measure of the effect of each of the exogenous variables on the rate of intrafirm diffusion is the elasticity of the time interval in Table 9.2 (between the dates when a firm was 10 percent and 90 percent dieselized) with respect to the exogenous variable. The estimated elasticities are $-0.35(\pi_i)$, $-0.60(L_i)$, $0.07(S_i)$, and $-0.49(C_i)$. All are evaluated at the means of the exogenous variables. These results would seem to indicate that the intrafirm rate of diffusion is most sensitive (in an elasticity sense) to changes in L_i and least sensitive to changes in S_i.[27]

[26] Berkson's weights are applied. See J. Berkson, "A Statistically Precise and Relatively Simple Method of Estimating the Bio-Assay with Quantal Response, Based on the Logistic Function," *Journal of the American Statistical Association*, Sept. 1953. Only those values of t such that $0.01 \leq P_i(t) < 1$ are included, and t is measured in years from 1939.

[27] According to the model, the interval in Table 9.2 equals $4.39/V_i$, which is approximately $4.39\,[c_1 + c_2\pi_i + c_3L_i + c_4S_i + c_5C_i]^{-1}$. Thus, when I represents this interval,

$$\frac{\delta I}{\delta \pi_i} \cdot \frac{\pi_i}{I} = -c_2\pi_i[c_1 + c_2\pi_i + c_3L_i + c_4S_i + c_5C_i]^{-1}.$$

Similar results are obtained for L_i, S_i, and C_i. Evaluating the results at the means of the exogenous variables, we get the numbers in the text.

5. Effect of Additional Factors

Some of the important factors that help to account for the unexplained variation in $\hat{V}_i$ seem fairly obvious. The intrafirm rate of diffusion was probably affected by the amount of pressure exerted on the firm by the diesel locomotive manufacturers. It was probably affected too by the training and preference regarding risk of the firm's technical officers and top management. Moreover, changes over time in the profitability of the firm's investment in diesel locomotives undoubtedly was a significant factor. Although these factors were probably important, they were omitted because no satisfactory way could be found to measure them.[28]

In this section, we investigate the effects of four additional variables that may have been important and that can be measured at least roughly. The first factor is the age distribution of the steam locomotives owned by the ith firm when it "began" to dieselize. When it is assumed that this age distribution was rectangular, with its upper end point at the replacement age, the percent of a firm's locomotives that had to be replaced each year after it "began" to dieselize is found to be inversely related to the range of this distribution. Moreover, the range is inversely related to M_i—the percent of the ith firm's steam locomotives that were 15 years old or more when it "began" to dieselize. Thus, since the intrafirm rate of diffusion would be expected to vary directly with the percent of a firm's steam locomotives that were due for replacement each year after it "began" to dieselize, one would expect $\hat{V}_i$ to be directly related to M_i. To check this, M_i was included as an additional independent variable in equation (12), the result being

$$\text{(13)} \quad \hat{V}_i = \underset{}{-0.257} + \underset{(0.515)}{0.849\,\pi_i} + \underset{(0.009)}{0.051\,L_i} - \underset{(0.0023)}{0.0030\,S_i} + \underset{(0.041)}{0.117\,C_i} + \underset{(0.0025)}{0.0011\,M_i}.$$

The regression coefficient for M_i has the right sign but is statistically nonsignificant.[29]

[28] Of course, another factor that might be important here is a firm's rate of growth—because of its impact on the extent of a firm's investment in new locomotives and the rate at which old locomotives are scrapped. Although we assume throughout that every firm's output remains the same for the duration of the period, this is only a convenient first approximation. Another factor that may have been important is reciprocity. Salter describes a number of factors influencing the rate of diffusion, but his analysis cannot easily be applied here because the railroad industry is regulated. His analysis is concerned primarily with free markets. (See W. Salter, *Productivity and Technical Change* [Cambridge, 1960].) Another relevant publication is G. Terborgh, *Dynamic Equipment Policy* (New York: McGraw-Hill, 1949).

[29] We assume that each firm's steam locomotives were replaced when they were X years old (the diesel being available as an alternative) and that the age distribution of each firm's steam locomotives was rectangular when it "began" to dieselize, the upper end point of the dis-

The second factor is A_i, the absolute number of diesel locomotives that the ith firm had to acquire in order to go from 10 to 90 percent of full dieselization. As this number increases, the firm is forced to invest more heavily each year in diesel locomotives in order to make this transition in a given length of time (and hence to maintain a given value of V_i). Since it may not be possible or worthwhile for firms of given size to support an annual investment exceeding some maximum amount, V_i may be inversely related to this number. To check this, A_i was included in equation (12), the result being

$$\text{(14)} \quad \hat{V}_i = \underset{}{-0.174} + \underset{(0.490)}{1.036\ \pi_i} + \underset{(0.008)}{0.047\ L_i} + \underset{(0.0118)}{0.0126\ S_i} + \underset{(0.040)}{0.108\ C_i} - \underset{(0.00028)}{0.00038\ A_i}.$$

The regression coefficient for A_i has the right sign but is statistically nonsignificant.[30]

The third factor is the average length of haul of the ith firm. Diesel locomotives would be expected to be particularly profitable for railroads that made long hauls, because intermediate service points could be eliminated. Thus, one might expect V_i to be directly related to Q_i—the ith firm's average length of haul (in miles) during 1937–1946. But when this variable is added to equation (12), we find once again that although the sign of its regression coefficient is "right," the coefficient is not statistically significant. More specifically, the result is

---

tribution being X. If so, the lower end-point must equal $X - (X - 15)/M_i$, and the range of the distribution (a measure of the amount of variation) must equal $(X - 15)/M_i$. Moreover, since the proportion of a firm's steam locomotives that had to be replaced each year immediately after it "began" to dieselize is the reciprocal of the range of the distribution, this proportion is directly related to M_i. Finally, one would expect the intrafirm rate of diffusion to be directly related to this proportion because, as the proportion of steam locomotives falling due for replacement each year increases, the minimum time required to attain full dieselization decreases. Of course, the assumption that the age distribution can be approximated by a rectangular distribution and that its maximum was X (which was the same for all firms) is very rough. Moreover, the minimum and actual time required to attain full dieselization may not be very closely related. These factors may explain the nonsignificant results in equation (13). Moreover, it is possible that the results would have been different if some cutoff point other than 15 years had been used. (The form of the basic data required that we use a measure like M_i to estimate the range.) Certainly, the present analysis of the effect of the age distribution of steam locomotives on the intrafirm rate of diffusion is only a beginning. Finally, we fit this factor into the model simply by adding M_i to the list of variables on the right-hand side of equation (3) and treating it like the other variables in subsequent equations. The result is that $\hat{V}_i$ should be a linear function of it too. (The same procedure is used in dealing with the other factors in this section.) But under some circumstances, it is difficult to fit M_i into this framework because its effects on $W_i(t)$ are likely to change with time after a while. This problem does not arise so acutely with the other three factors discussed in this section.

[30] Note that, when A_i is included in the regression, the regression coefficient for S_i be-becomes positive. Of course, this is because A_i and S_i are highly correlated.

$$(15) \quad \hat{V}_i = \underset{}{-0.268} + \underset{(0.500)}{1.058\ \pi_i} + \underset{(0.008)}{0.052\ L_i} - \underset{(0.0025)}{0.0043\ S_i} + \underset{(0.040)}{0.106\ C_i} + \underset{(0.00031)}{0.00039\ Q_i}.$$

The fourth factor is the profitability of the ith firm. More profitable firms might be expected to have higher values of V_i because they were better able than other firms to finance the necessary investment and to take risks. To check this, K_i—the average railway operating income as a percent of its total adjusted capital in the 2 years prior to and including the year when it "began" to dieselize—is included as an additional independent variable in equation (12), the result being

$$(16) \quad \hat{V}_i = -0.087 + \underset{(0.520)}{0.692\ \pi_i} + \underset{(0.008)}{0.052\ L_i} - \underset{(0.0023)}{0.0032\ S_i} + \underset{(0.040)}{0.113\ C_i} - \underset{(0.0073)}{0.0086\ K_i}.$$

The regression coefficient for K_i has the wrong sign and is statistically nonsignificant.[31]

Thus, although M_i, A_i, Q_i, and K_i might be expected to influence V_i in the ways we describe, there is no real evidence that they exerted such an influence. Their apparent effect is almost always in the expected direction, but it is statistically nonsignificant in every case.

6. Results Based on Utilization Data for Freight Service

Our measure of the rate of intrafirm diffusion is based on the number of locomotive units of each type (steam and diesel) owned by a firm. This measure suffers from the fact that locomotives differ in size and capacity and that some locomotives may be used little (if at all). It also suffers from the fact that freight, passenger, and switching services have to be lumped together. It would be preferable to use a measure based on the growth over time in the percent of total work done by diesels and to separate various types of services, but only a small amount of data of this sort has been published.

Using the available data regarding freight service (probably the most significant type of work), we try to determine whether our findings would have been modified substantially if utilization data of this sort, rather than ownership data, had been used. The model in section 3 can easily be modified to accommodate such data. Let N'_i be the total freight ton-miles of the ith

[31] Of course, this result may be due to inadequacies in the data regarding K_i. For example, the time periods to which these data pertain are somewhat arbitrary. Note too that this result is quite consistent with Eisner's findings regarding the investment function. See R. Eisner, "A Distributed Lag Investment Function," *Econometrica* (Jan. 1960).

firm during the relevant period, $D'_i(t)$ be the number of freight ton-miles hauled by diesels at time t, and

$$W'_i(t) = [D'_i(t+1) - D'_i(t)]/[N'_i - D_i(t)]$$
$$= g[\pi_i, U_i(t), S_i, C_i, \ldots].$$

Proceeding as we did in section 3 and letting $P'_i(t)$ be the proportion of the ith firm's total freight ton-miles hauled by diesels at time t, we find that

$$P'_i(t) = [1 + e^{-(\beta_i + H_i t)}]^{-1}$$
$$H_i = d_1 + d_2\pi_i + d_3L_i + d_4S_i + d_5C_i + e''_i,$$

where e''_i is an error term.

From published ICC data, it is possible to piece together enough information regarding $P'_i(t)$ to allow a rough test of this model for twenty of the firms in Table 9.3. Apparently a logistic function provides a reasonably good fit. Using the rough estimates of H_i shown in Table 9.4 (and omitting e''_i), we find that

$$\text{(17)} \quad \hat{H}_i = \underset{}{-0.268} + \underset{(1.732)}{0.502\,\pi_i} + \underset{(0.040)}{0.138\,L_i} - \underset{(0.007)}{0.016\,S_i} + \underset{(0.114)}{0.391\,C_i},$$

the coefficient of correlation being 0.83. When M_i, A_i, Q_i, and K_i are inserted into equation (17) as additional independent variables, the results are generally like those in the previous section, the only notable difference being that the effect of M_i is statistically significant.[32]

Thus, these fragmentary data regarding the utilization of diesels in freight service yield the same general kind of results as those obtained from ownership data. The effects of π_i, L_i, S_i, C_i, and M_i are in the same direction, and the model in section 3 seems to fit about as well in one case as in the other.

[32] For 1951–1954, we could obtain the percent of freight ton-miles hauled by diesel locomotives for twenty-three of the firms in Table 9.4. These figures came from the ICC's *Monthly Comment on Transport Statistics* (1951, 1952, and 1953) and *Moody's*. Prior to 1951 we could obtain the percent of freight locomotive miles accounted for by diesel locomotives for all the firms. These figures came from the ICC's *Comparative Statement of Railway Operating Statistics*. Ignoring the differences between these two measures, we computed the time interval between the date when the percentage equaled 10 and the date when it equaled 90. Then we divided this interval into 4.39 to obtain a rough estimate of H_i. Such an estimate could be made for only twenty firms, since not all firms had reached 90 percent by 1954. See Table 9.4. These seem to be the only published data that are available. Of course, to be consistent, we should probably have based our measure of L_i, C_i, M_i, and K_i on the date when diesel locomotives accounted for 10 percent of freight ton-miles rather than 10 percent of the locomotive stock. But it is doubtful that this would have made any appreciable difference. When M_i is added to equation (17), the result is

$$\hat{H}_i = -1.76 - \underset{(1.572)}{0.114\,\pi_i} + \underset{(0.036)}{0.136\,L_i} - \underset{(0.006)}{0.017\,S_i} + \underset{(0.100)}{0.395\,C_i} + \underset{(0.008)}{0.018\,M_i}.$$

The correlation coefficient is almost 0.9.

The difference is that some coefficients are statistically significant in one case but not in the other, the effects of M_i and S_i being significant here (but not in sections 4 and 5), and the effect of π_i being nonsignificant here (but significant in sections 4 and 5).

7. Summary and Conclusions

The principal conclusions of this chapter are as follows: First, we constructed a simple model to explain how rapidly, once a firm began to dieselize, it substituted diesel motive power for steam. When tested against data for thirty Class I railroads, this model seemed to stand up quite well. About 70 percent of the interfirm variation in the rate of dieselization could be explained, and the effect of each exogenous variable was in the expected direction and (with one exception)[33] statistically significant. Although the model is obviously highly simplified and incomplete, it is of considerable help in explaining the substantial differences among the intrafirm rates of diffusion of this innovation.

Second, since these findings pertain to only one innovation, they provide little information regarding the usefulness of a model of this sort for new techniques in general. However, from what little additional evidence we have, there seems to be a good chance that the same sort of model would be useful in dealing with a wide class of innovations.[34] If so, this would have at least four implications: (1) It would mean that the same kind of model can be used to represent both the rate of diffusion among firms and the rate of diffusion within a firm. The model used here emphasizes the same sorts of explanatory factors and is similar in structure to one used with considerable success in Chapter 7 to represent the rate of interfirm diffusion of an innovation. The fact that the same sort of model works reasonably well in both cases suggests that there is a considerable amount of unity and similarity between the two diffusion processes. Moreover, the results in each case lend support to those in the other case.

(2) Together with previous results, it would suggest that there exists an important economic analogue to the classic psychological laws relating reaction time to the intensity of the stimulus.[35] The profitability of an investment opportunity acts as a stimulus, the intensity of which seems to govern quite closely a firm's speed of response. In terms of the diffusion process, it governs

[33] See note 36.

[34] As we pointed out in note 21, the results from the studies previously made of the diffusion process seem to indicate that the factors considered here are important (and that they operate in the expected direction) for innovations of various sorts in a variety of industries.

[35] See, for example, J. Cattell, "The Influence of the Intensity of the Stimulus on the Length of the Reaction Time," reprinted in Dennis, *Readings in the History of Psychology* (New York: Appleton-Century-Crofts, 1948). Note that π_i has a statistically significant effect only when the ownership rather than the utilization data are used. However, if utilization data for all thirty firms had been available, I strongly suspect that it would have had a significant effect in both cases. Moreover, M_i, which has a significant effect on Π_i, is a partial measure of the profitability of the innovation.

both how rapidly a firm begins using an innovation and how rapidly it substitutes it for older methods.

(3) If the effect of a firm's size is generally like that found here, it would be of considerable interest to economists concerned with problems regarding industrial concentration and the large firm. In line with the allegations of Stocking [161], Yance [169], and others, it would appear that small firms, once they begin, are at least as quick to substitute new techniques for old as their larger rivals. Although this is obviously only one of a great many considerations in formulating policy in this area, it is worthy of attention.[36]

(4) The results point up the importance in this regard of the time when a firm begins to use the innovation, the age of its equipment at that time, and its liquidity. All of these factors have a statistically significant effect on the intrafirm rate of diffusion (measured in terms of either the ownership or the utilization data or both).[37] However, as so often has been the case in studies of investment behavior, the effect of the profitability of the firm is not statistically significant.

[36] Note that S_i has a statistically significant effect only when the utilization, rather than the ownership, data are used. Previous chapters have touched on various other respects of the relationship between firm size, market structure, and technical progressiveness.

[37] M_i has a statistically significant effect only when the utilization, rather than the ownership, data are used.

Part V

Conclusion

CHAPTER 10

Summary of Findings

Finally, we review very briefly the nature and significance of our findings. This chapter brings the results of previous chapters to bear on the following questions, which are undoubtedly some of the most important in this area: First, what determines the rate of technological change? Second, how much research and development of various kinds is performed in the American economy? Third, what determines the level of industrial research and development expenditures? Fourth, what is the relationship between the level of such expenditures and various rough measures of inventive output? Fifth, how are a firm's total expenditures on research and development allocated among projects, and what are the characteristics of the projects that are undertaken? Sixth, to what extent can we estimate the returns from industrial research and development expenditures? Seventh, what is the role of the very large firm in promoting technological change and the utilization of new techniques? Eighth, when do important innovations take place and what are their effects on the innovator's growth rate? Ninth, what determines how rapidly innovations spread? Tenth, what are the characteristics of technical leaders and followers?

1. The Rate of Technological Change

What determines the rate of technological change? Only recently have economists begun to give this question the attention it deserves, the result being that existing theory is still in a relatively primitive state.[1] On a priori grounds, one would expect an industry's rate of technological change to depend on the amount of resources devoted by members of the industry and by the government to the improvement of the industry's technology. In addition, it depends on the effectiveness with which the resources are used, as well as on the quantity of private and public resources devoted to the improvement of technology in other industries, particularly those that supply relevant equipment, components, and materials. The amount of resources devoted by the government to the improvement of an industry's technology depends on how closely the industry is related to the defense, medical, and other social needs for which the government assumes major responsibility; on the extent

[1] The common measures of the rate of technological change have important shortcomings. Besides the effects of "pure" technological change, they contain the effects of whatever other factors are not explicitly included. In addition, there are other problems which are discussed in E. Mansfield *The Economics of Technological Change* (New York: Norton, 1968), Chap. 2.

of the external economies generated by the relevant research and development; and on more purely political factors. The amount of resources devoted by private industry depends heavily on the profitability of their use.

Accepting the proposition that the amount invested by private sources in improving an industry's technology is influenced by the anticipated profitability of the investment, it follows that the rate of technological change in a particular area is influenced by the same kinds of factors that determine the output of any good or service. On the one hand, there are demand factors which influence the rewards from particular kinds of technological change. For example, if a prospective change in technology reduces the cost of a particular product, increases in the demand for the product are likely to increase the returns from effecting this technological change. Similarly, a growing shortage and a rising price of the inputs saved by the technological change are likely to increase the returns from effecting it. On the other hand, there are also supply factors which influence the cost of making particular kinds of technological change. Obviously, whether people try to solve a given problem depends on whether they think it can be solved and how costly it will be, as well as on the payoff if they are successful. The cost of making science-based technological changes depends on the number of scientists and engineers in relevant fields and on advances in basic science; for example, advances in physics reduced the cost of effecting changes in technology in the field of atomic energy. In addition, the rate of technological change depends on the amount of effort devoted to making modest improvements that lean heavily on practical experience.

Of course, these factors are not the only ones that influence the rate of technological change. As emphasized in various chapters, there is considerable uncertainty in the research and inventive processes, and laboratories, scientists, and inventors are motivated by many factors other than profit. Also, the rate of technological change is influenced by the industry's market structure; the legal arrangements under which it operates; the attitudes toward technological change of management, workers, and the public; the way in which the firms in the industry organize and manage their research and development; the way in which the scientific and technological activities of relevant government agencies are organized and managed; and the amount and character of the research and development carried out in the universities and in other countries.[2]

2. Research and Development Expenditures in the United States

How much is spent on research and development? According to the National Science Foundation, the total equaled $17 billion in 1963. There has been a very great increase during the past several decades in the amount spent on research and development. This is true regardless of whether one

[2] For a more complete discussion, see Mansfield, *Economics of Technological Change*, Chap. 2.

considers only the total expenditures or whether one looks separately at the amount spent by government, industry, universities, or other nonprofit institutions. In each of these sectors, expenditures on research and development have increased enormously. Moreover, if total expenditures are broken down by industry, the results are the same. In every industry, research and development expenditures have increased in absolute terms and as a percent of sales.

Although industry does the lion's share of the work,[3] the federal government finances most of the research and development that industry performs—and most of the research and development performed by the universities and the other nonprofit organizations as well. The percentage of industrial, university, and other nonprofit organizations' research and development financed by the federal government has increased considerably since World War II. Three departments and agencies of the federal government account for nearly 90 percent of its expenditures on research and development. They are the Department of Defense, the National Aeronautics and Space Administration, and the Atomic Energy Commission. The primary purpose of the expenditures made by these agencies is to develop and improve weapons systems, to push forward the nation's space program, and to develop new applications for atomic energy. Obviously, the research and development expenditures of the federal government are intimately linked with the cold war.

Private industry, although somewhat overshadowed in this respect by the federal government, also finances an enormous amount of research and development—about $5.6 billion in 1963. Indeed, most of the research and development carried out in industries other than aircraft, electrical equipment, and instruments is financed privately. As a percent of sales, company-financed research and development expenditures are particularly high in the instrument, electrical equipment, chemical, and aircraft industries. When the full range of firm size is considered, there is a tendency in most industries for the percent of sales spent on research and development to increase with the size of the firm.[4]

3. Industrial Research and Development Expenditures

What determines how much company-financed research and development a firm carries out? According to the simple model set forth in Chapter 2, a firm sets its research and development expenditures so as to move part way from the previous year's level toward a desired level that depends on the firm's expectation regarding the average profitability of the research and development projects at hand, the profitability of alternative uses of its funds, and its size. The firm's speed of adjustment toward the desired level depends

[3] Industry does most of the R and D, but if one considers basic research alone, the universities and other nonprofit institutions do most of the work.

[4] This statement is based on a rather coarse classification of firms by size, and it tells little or nothing about the differences in this regard between the largest firms and their somewhat smaller competitors. These differences are discussed in section 3.

on the extent to which the desired level differs from the previous year's level and on the percent of its profits spent during the previous year on research and development. This model was formulated in part on the basis of interviews with research directors and other executives of a number of firms in the chemical and petroleum industries.

For eight firms where the necessary data could be obtained, this model, in more specific and operational form, could fit historical data regarding these firms' expenditures quite well. Moreover, when supplemented with additional assumptions, it could fit the 1945–1958 data for thirty-five firms in five industries (petroleum, chemicals, drugs, glass, and steel) quite well, and it could do a reasonably good job of "forecasting" their 1959 expenditures. Of course, the model is a more apt description of decision-making regarding applied research and development than basic research, but the latter is quite small in this context.

Because of the small number of observations and the roughness of the basic data, it should be stressed that the results are tentative. But if reasonably trustworthy, they have at least three significant implications. First, they allow us to make rough estimates of the effect of certain kinds of government policies on the amount a firm spends, in money terms, on research and development. These estimates are interesting from an exploratory viewpoint but it would probably be unwise to attach much policy significance to them at present.

Second, the fact that the model fits the data so well seems to imply that the process by which a firm's research and development expenditures are determined is not so divorced from profit considerations as some observers have claimed. If firms "establish research laboratories without any clearly defined idea of what the laboratories could perform" [125], and blindly devote some arbitrarily determined percentage of sales to research and development, it is difficult to see why the model fits so well.

Third, the results provide new evidence regarding the effects of a firm's size on the amount spent on research and development. They indicate that among large- and medium-sized firms in the petroleum, chemical, glass, steel, and drug industries there was no tendency, except in the chemical industry, for the percent of a firm's sales devoted to research and development to increase with the size of the firm. If anything, the opposite was the case.

4. Research and Development and Inventive Output

The productivity of industrial research and development is an extremely important variable which is plagued by unusually difficult measurement problems. On the basis of the crude measurements that can be made, does it seem that a firm's output of significant inventions is closely related to the amount it spends on research and development? Is there any evidence that the productivity of a firm's research and development activities increases with the amount spent on research and development? Is there any evidence that productivity is greater in the largest firms than in somewhat smaller ones?

To help answer these questions, studies were made, as reported in Chapter 2, of the number of significant inventions carried out by various large firms in the chemical, petroleum, and steel industries. Calculations based on these crude data suggest the following three conclusions: First, when the size of the firm is held constant, the number of significant inventions carried out by the firm seems to be highly correlated with the size of its research and development expenditures. Thus, although the pay-out from an individual research and development project is obviously very uncertain, it seems that there is a close relationship over the long run between the amount a firm spends on research and development and the total number of important inventions it produces.

Second, the evidence from this cross-section analysis seems to suggest that increases in research and development expenditures, in the relevant range and when the size of the firm is held constant, result in more than proportional increases in inventive output in chemicals. But in petroleum and steel, there is no real indication of either economies or diseconomies of scale within the relevant range. Thus, except for chemicals, the results do not indicate any marked advantage of very large-scale research activities over medium-sized and large ones.

Third, when a firm's expenditures on research and development are held constant, increases in size of firm seem to be associated in most industries with decreases in inventive output. Thus, the evidence suggests that the productivity of a research and development effort of given scale is lower in the largest firms than in the medium-sized and large ones.

5. The Allocation and Characteristics of the Firm's Research and Development Portfolio

The allocation of the firm's funds among research and development projects, the characteristics of the projects that are undertaken, and the probable outcome of these projects were explored in the detailed case study presented in Chapter 3. This study investigated the research and development portfolio of the central research laboratory of one of the nation's largest firms. Although the evidence is limited, the assumption of expected profit maximization seemed to be of use in explaining the allocation of funds in the laboratory (excluding basic research). About half of the variation in the allocation of funds could be explained by a model which assumes that proposed spending is increased to the point at which the increase in the probability of success is no longer worth its cost.

Although expected profit maximization could explain about half of the variation in the allocation of funds, about half remained unexplained. Four factors seemed to account for much of this unexplained variation. (1) When expected profit is held constant, safe projects are preferred over risky ones. (2) Some attempt is made to satisfy scientific as well as commercial objectives, the consequence being that some projects are justified more on the basis of scientific interest than on expected profit. (3) Intrafirm politics are important; for example, projects differ considerably in the amount of pressure applied by operating executives to have them carried out. (4) Some scientists and department managers are much more effective than others in arguing for their proposals and in mobilizing support for them.

A detailed description of the laboratory's applied research and development projects seemed to indicate that most are expected to be completed in 4 years or less, and the results are expected to be applied only a few months later. The estimated probability of technical success averages about 0.80; and if the projects are successful, the estimated rate of return from the investment in research and development is extremely high. These results are not very different from some preliminary findings pertaining to several large firms in the chemical and petroleum industries.

Failure rates and slippages in schedule, which are used repeatedly to measure the extent of the uncertainty in research and development, seem to exaggerate very greatly the extent of the technical uncertainties involved. About one-half of this laboratory's projects did not achieve their technical objectives on time. However, about two-thirds of the "failures" resulted from changes in objective or the transfer of personnel to other projects. In only about one-third of these cases was there any evidence that "failure" was due to technical difficulties. Thus, whereas the unadjusted failure rate would indicate that the average probability of technical success was about 0.50, it was really about 0.75.[5]

A comparison of the estimate of the probability of a project's technical success (made prior to the beginning of the project) with the outcome of the project indicates that such estimates have some predictive value, but not a great deal. A discriminant function based on these estimates predicted incorrectly in about one-third of the cases. This finding should be useful in the formulation and evaluation of various research and development budget allocation techniques, practically all of which rely on such estimates. Up to this point, no information was available regarding their accuracy.

6. Rates of Return from Industrial Research and Development

Building on the conventional economic theory of production, is it possible to derive a technique to estimate the marginal rate of return from a firm's research and development expenditures? If the production function is Cobb-Douglas, if total past research and development expenditures, as well as labor and capital, are inputs, and if research and development expenditures have grown exponentially, one can obtain relatively simple expressions for

[5] Note that this is the probability of technical, not commercial, success. Of course, the probability of both technical and commercial success would be lower than this.

the marginal rate of return from research and development, whether technological change is capital-embodied or disembodied. If it is capital-embodied, the marginal rate of return is directly related to the elasticity of output with respect to total past research and development expenditures and the rate of investment, but inversely related to the amount spent in the past on research and development and the capital-output ratio. If it is disembodied, the marginal rate of return is directly related to the elasticity of output with respect to total past research and development expenditures and inversely related to the ratio of total past research and development expenditures to present output.

On the basis of these theoretical results, presented in Chapter 4, estimates of the marginal rates of return in 1960 were made for ten major chemical and petroleum firms, and lower-bounds for the marginal rates of return were estimated for ten manufacturing industries. From the data for individual firms, it appears that the rate of return was very high in petroleum; in chemicals, it was high if technological change was capital-embodied but low if it was disembodied. The rate of return was directly related to a firm's size in chemicals, but inversely related to it in petroleum. From the industry data, it appears that the rate of return was relatively high (15 percent or more) in the food, apparel, and furniture industries. These findings are of interest, but not much policy significance can be attached to them, for reasons discussed below.

The rate of technological change, both at the level of the firm and the industry, is directly related to the rate of growth of its accumulated research and development expenditures. On the firm level, this would be expected on the basis of our model. However, there is no evidence that such frequently used variables as the industry's (or firm's) ratio of research and development expenditures to sales, its rate of growth, or its concentration ratio exert an important influence on its rate of technological change—apart from their possible influence on the rate of growth of its accumulated research and development expenditures. These findings are based on small samples; much more data and better measurements are needed.

There are enormous difficulties, both conceptual and practical, in estimating the returns from industrial research and development. These results should be viewed with considerable caution, for at least four reasons. First, they are based on a number of highly simplified assumptions—that uncertainty can be ignored, that technological change is cost-reducing, that all technological change is neutral, that the production function is Cobb-Douglas, and that capital is "putty," not "hard-baked clay." Second, they contain substantial sampling errors. Third, they are incomplete estimates of the social rate of return. Fourth, although it is easy to include lags in the effect of research and development expenditures on the production function, as well as a finite elasticity of supply of research and development inputs to the firm, this was not done because of the lack of relevant data.

7. The Size of Innovators

The significance of the size of the innovator, the firm responsible for the first commercial application of an invention, has been stressed repeatedly by economists. Is it true that the largest firms have been the first to introduce a disproportionately large number of the important new processes and products that have been developed in recent years? Is it true that they dominate the picture to a larger extent now than in the past? To help answer these questions, studies were made of the iron and steel, petroleum, bituminous coal, and railroad industries to determine whether in each case the largest four firms seemed to introduce a disproportionately large share of these innovations. Then a simple model was constructed to explain why the giant firms accounted for a disproportionately large share of the innovations in some cases, but not in others; and an attempt was made to estimate whether innovations would have been introduced more slowly if these large firms had been broken up.

The principal results are as follows: First, although it is often alleged that the largest firms do more than their share of the pioneering, this is not always the case. For example, in petroleum refining, bituminous coal, and railroads, the largest firms accounted for a larger share of the innovations than they did of the market, but in iron and steel, they accounted for a smaller share. Second, the evidence seems quite consistent with a simple model which predicts that the largest four firms will do a disproportionately large share of the innovating in cases where (1) the investment required to innovate is large relative to the size of the firms that could use the innovation, (2) the minimum size of the firm required to use the innovation is large relative to the average size of the firm in the industry, and (3) the average size of the largest four firms is much greater than the average size of all potential users of the innovation. Third, there is some evidence in the petroleum, bituminous coal, and steel industries that the largest few firms carried out no more innovations, relative to their size, than did somewhat smaller firms (considerably smaller ones in steel). However, these estimates are very rough. Fourth, there is evidence that the smallest steel, bituminous coal, and oil firms did less innovating during 1939–1958—relative to large- and medium-sized firms—than in 1919–1938. With the rising costs of development and the greater complexity of technology, this is not surprising.

8. The Timing of Innovation

Some important questions regarding innovations are concerned with their timing and their effect on the level and timing of investment in plant and equipment. The available evidence indicates that the average lag between invention and innovation is about 10–15 years. For petroleum innovations, the standard deviation of this lag is about 5 years; in all other industries combined, it is about 16 years. Apparently, mechanical innovations require the shortest interval, and electronic innovations require the longest. The lag seems shorter for consumer products than for industrial products, and shorter for innovations developed with government funds than for those developed with private funds.

To see whether the rate of occurrence of major innovations varies appreciably over the business cycle, the dates of first commercial introduction were determined for 150 processes and products introduced during 1919–1958 in the iron and steel, petroleum-refining, and bituminous coal industries. On the basis of these data, analyzed in Chapter 6, it appears that process innovations were most likely to be introduced during periods when the industries were operating at about 75 percent of capacity. Contrary to the opinion of many economists, there was no tendency for process innovations to cluster during the periods when operating rates were extremely high or extremely low. Apparently, innovation at the trough was discouraged by the meagerness of profits and uncertainty regarding the future. At the peak, some executives in the industries claim that it was discouraged by the lack of unutilized capacity where alterations could be made cheaply and without interfering with production schedules. For product innovations, there was no evidence that the rate of innovation varied significantly over the business cycle.

An investment function combining the flexible capacity accelerator with a simple model of innovation-induced investment can explain the behavior of the level of investment in steel and petroleum more adequately than the accelerator alone. The timing of innovation seems to have had a significant effect on the level and timing of expenditures on plant and equipment. However, it is a difficult variable to handle empirically and these results are presented merely as the findings of a crude experiment, not of a definitive study.

9. Innovation and the Growth of Firms

How large has been the payoff for a successful innovation? Perhaps the best single measure of a firm's rewards is the rate of return on its investment; but because of data limitations, we investigated in Chapter 6 the effect of a successful innovation on a firm's growth rate, another interesting, if incomplete, measure of its success. A comparison of the growth rates of the successful innovators—during the period in which the innovation occurred—with those of other firms of comparable initial size helps to indicate how great the payoff was, in terms of growth, for a successful innovation. A comparison of the preinnovation and postinnovation growth rates of the innovators also provides evidence on this score.

In every time interval, the successful innovators in steel and petroleum grew more rapidly (during a 5- to 10-year period after the innovation occurred) than the other firms, their average growth rate often being more than twice that of the others. Moreover, in the period after they introduced the innovations, the difference in growth rate between innovators and other comparable firms was greater than before the introduction of the innovation. According to our best estimates, the average effect of a successful innovation was to raise a firm's annual growth rate by 4 to 13 percentage points, depending on the time interval and the industry. When each innovator is considered separately, the difference between its growth rate and the average growth rate of other comparable firms seems to have been inversely related to its size. As one would expect, a successful innovation had a much greater impact on a small firm's growth rate than on a large firm's.

10. Rates of Diffusion of Innovations

The origin and initial introduction of an innovation having been considered, it is necessary to look next at the subsequent diffusion process. Once an innovation has been introduced by one firm, what factors determine the rate at which other firms follow the innovator? What determines the intrafirm rate of diffusion of an innovation?

An intensive study, reported in Chapters 7 and 9, of the diffusion of a dozen major process innovations in the railroad, brewing, steel, and bituminous coal industries seems to indicate the following answers to these questions. First, the diffusion of a major new technique is a fairly slow process. Measured from the date of first commercial application, it often took 20 years or more for all the major firms in the industry to install an innovation. Seldom did it take much less than 10 years. When one innovation is compared with another, there are very marked differences in the rates of diffusion, some taking much longer than others to spread throughout an industry.

Second, there seems to be a definite "bandwagon" or "contagion" effect. As the number of firms in an industry using an innovation increases, the probability of its adoption by a nonuser increases. This is because, as experience and information regarding an innovation accumulate, the risks associated with its introduction grow less and competitive pressures mount. Moreover, in cases in which the profitability of an innovation is difficult to assess, the mere fact that a large proportion of a firm's competitors have adopted the innovation may prompt the firm to consider it more seriously.

Third, the rate of diffusion tends to be higher for more profitable innovations and for those requiring relatively small investments. The rate of diffusion also differs among industries, there being some slight indication that it is higher in less concentrated industrial categories. The relationship between these variables and the rate of diffusion is in accord with a simple mathematical model of the imitation process, and is surprisingly close. This model may be useful for forecasting purposes, although one should note that in none of the cases considered were patents a relevant factor, since they were held by the equipment producers.

Fourth, a study of the intrafirm rate of diffusion of diesel locomotives suggests that there are great differences among firms in the rate at which they substitute an innovation for an older method. In the case of the diesel locomotive, a substantial part of this variation can be explained by interfirm differences in the profitability of investing in diesel locomotives, interfirm differences in size and liquidity, and interfirm differences in the date when the process of dieselization began. Increases in each of these factors other than firm size result in increases in the intrafirm rate of diffusion.

11. Characteristics of Leaders and Followers

What factors seem to determine whether one firm will be quicker than another to begin using a particular technique? Do the same members of an industry tend to lag behind in introducing innovations, or are the leaders in one case likely to be the followers in another? According to the results shown in Chapter 8, the speed with which a particular firm begins using a new technique is directly related to the firm's size and the profitability of its investment in the technique. For example, in the cases considered, if one firm is four times as large as another, the profitability of the investment in the innovation being the same for both, the probability that it will introduce an innovation more rapidly than its smaller competitor is about 0.80. Similarly, if the innovation is considerably more profitable for one firm than another (of equal size), the probability is generally quite high that the former firm will be quicker to introduce it.

A firm's financial health, as measured by its profitability, liquidity, and growth rate, bears no close relationship to how long it waits before introducing a new technique. Whereas some relatively prosperous members of an industry tend to be quick to introduce an innovation, others tend to be slow. If the size of the firm and the profitability of the innovation are held constant, the relationship seems to be quite weak, if existent at all. Perhaps these variables are less important than other more elusive and essentially noneconomic variables. The personality attributes, interest, training, and other characteristics of top and middle management may play a very important role in determining how quickly a firm introduces an innovation. The presence or absence of a few men in the right places who believe strongly in the value of the new technique may make a crucial difference.

Our studies show quite clearly the dangers involved in the common assumption that certain firms are repeatedly the leaders, or followers, in introducing new techniques. It would be very misleading to assume, from just a few innovations, that the firms that are quick to use them are generally the leaders in this sense. According to our findings, there is a very good chance that these firms will be relatively slow to introduce the next innovation that comes along.

12. Implications for Policy[6]

This book is primarily a report of basic research. It is important that these findings be treated with caution and that the user be willing to resist the temptation to build policy implications on theoretical and empirical foundations that are too fragile to support them. Nonetheless, a few words should be added concerning areas of application of our findings. First, we have provided new results regarding the role of the large firm in promoting technological change and the utilization of new technology. These results do not indicate that total research and development expenditures (in most of the industries we studied) would decrease if the largest firms were replaced by somewhat smaller ones. They do not indicate that the research and development expenditures carried out by the largest firms are generally more productive than those carried out by somewhat smaller firms. They do not suggest that greater concentration results in a faster rate of diffusion of innovations. However, if innovations require a large amount of capital, they do suggest that the substitution of fewer large firms for a larger number of smaller ones may lead to quicker commercial introduction.

Second, we have provided new findings regarding the diffusion of new techniques. These results may help firms forecast the rate of growth of sales of new products. Given an estimate of the profitability of a new product to its potential users and the size of the investment required to adopt it, one can use the model in Chapter 7 to estimate the rate at which the number of users will grow. The results in Chapters 8 and 9 may be useful in predicting intrafirm rates of diffusion and the characteristics of the firms that are most likely to purchase the product first. Several firms have begun experimenting in this way with these results. In government applications, the results should also be of use to decision-makers interested in promoting the transfer of technology from military, space, and other government work to civilian uses. They provide some indication of the length of time that such transfer generally takes and the factors that seem to determine the extent of the lag. A number of our results and techniques have already been used for these and other purposes in studies sponsored by government agencies.

Third, we have provided new findings regarding the impact and outcome of industrial research and development. Some measures of the relationship between a firm's research and development expenditures and its inventive output, as well as of the effect of successful innovations on a firm's growth rate, have been provided. The crudeness of these estimates must be emphasized, but the results may be a useful first step toward the formulation of

[6] For a much more complete discussion of public and private policy questions in this area, see Mansfield, *Economics of Technological Change*, Chaps. 3–7. Also see Mansfield, "Technological Change: Measurement, Determinants, and Diffusion," *Report to the President by the National Commission on Technology, Automation, and Economic Progress* (1966).

operational techniques to solve some of the most important and difficult problems facing firms in this area. Also, firms may be interested in the results given in Chapter 3 regarding the causes of project failure, the slippages in schedule, and the accuracy of the estimates of the probability of success made before the projects were started. The accuracy of the estimated probabilities of success is particularly important, since the techniques recommended by economists and operations researchers for selecting research and development projects generally depend on the use of such estimates.

13. Limitations

In each of the previous chapters, an attempt was made to set forth the more important qualifications and limitations of the findings. Before concluding, some of these limitations should be noted once again. First, there is the matter of scope. I have made no attempt to cover in detail all aspects of the economics of technological change. The subject is so vast and the literature on some topics—e.g., the problems of adjusting to technological change—is so voluminous that it would be impossible to do so in a volume of this length. Rather than spread my attention over all aspects of the subject, I have looked in detail at a few of the more important questions regarding the process by which new processes and products are created and assimilated. A more comprehensive treatment of the field is provided in my *Economics of Technological Change*.[7]

Second, there is the matter of methodology. The mathematical models that are used are often very simple, in part because of necessity, in part because of choice. Although these models seem to be useful approximations, their roughness should be noted. Also, the data that could be obtained are sometimes limited in both quantity and quality. In some studies, we were forced to use rather small samples, because no published data were available and it was necessary to collect the data ourselves—a very time-consuming and expensive process.

Third, there is the matter of data coverage. Although some chapters are based on data covering most of the economy, others—because of the detailed nature of the analysis—must rely heavily on data pertaining to only a few industries. Considering these latter chapters as a whole, a fairly broad cross section of American industry is included—steel, petroleum, electrical equipment, electronics, coal, railroads, brewing, chemicals, drugs, and glass. However, in any particular chapter, the industrial coverage may be fairly narrow.

The studies contained here are tentative in a great many respects. They represent some of the first attempts to apply econometric techniques to the process of technological change, and they are part of a continuing program of research that I am carrying out in this area. No pretense is made that they are close to the last word on the subject. Hopefully, however, they represent some of the most advanced work that has yet been attempted.

[7] See E. Mansfield, *The Economics of Technological Change* (New York: Norton, 1968).

14. The Need for Additional Research

A book of this sort inevitably ends with a plea for more research—and in this respect I shall make no attempt to break with tradition. Despite the advances that have been made in the last decade, too little is known about industrial research and technological innovation in the American economy. Much more work is required.[8] Hopefully, this book will help to stimulate economists and others to turn their attention to these types of research. Although this has become a fashionable area, there is little danger of its being overemphasized in the near future.

[8] In a paper commissioned by the Carnegie Corporation of New York, I listed the types of research that, in my opinion, are most needed. See "The Economics of Research and Development: A Survey of Issues, Findings, and Needed Future Research," *Patents and Progress* (Homewood, Ill.: Irwin, 1965).

Appendices
References
Index

Appendix A

Estimates of $\tilde{R}_i(t)$, $\bar{\rho}_i(t)$, ρ^*, and $\theta_i(t)$: Interviews were obtained in 1960 with executives, ordinarily the president or research director, of three petroleum and three chemical firms. In 1963, interviews were obtained once again with executives of five of these firms. Numerous questions were asked, but three are of particular importance as sources of the data in Table 2.1. First, to determine $\tilde{R}_i(t)$ in 1958 and 1962, they were asked to estimate how much the firm would have spent in these years if it could have acquired instantaneously all of the personnel and equipment that it wanted at existing prices, and if it could have avoided the inefficiencies resulting from rapid changes in R and D expenditures. In a few cases, it was also possible to obtain such estimates for earlier years in the 1950's, these estimates of $\tilde{R}_i(t)$ being used in the "predictions" in Figure 2.2.

Second, they were asked to estimate the frequency distribution of projects (that would have been carried out in 1958 and 1962 under those circumstances) by expected profitability. These estimates generally allowed for differences among projects in risk, rough estimates of the probability of success being used by almost all firms in their measures of a project's potential profitability. The actual units in which these estimates were expressed differed from firm to firm, depending on the sorts of evaluation procedures they used. Note that when this question was asked the executives were not told what their replies would be used for. In particular, the question was not posed in such a way that they would make the answer correspond with the size of their R and D expenditures in order to make their actions seem "rational."

The lower-bound of the estimated frequency distribution is used as an estimate of ρ^*; the average of the distribution is used as an estimate of $\bar{\rho}_i(t)$. The estimates of ρ^* differ from firm to firm, but I assume that this is due largely to differences among firms in the units in which ρ^* is measured. In computing each firm's value of $\bar{\rho}_i(t)/[\rho^* - \bar{\rho}_i(t)]$, I used its own value of ρ^*, and assume that the results are equivalent to those that would have been obtained if the units of measurement were the same. This assumption seems reasonable.

Third, they were asked how much the planned, or budgeted, expenditures on R and D for 1958 and 1962 differed from the actual expenditures. Using this piece of information, it was possible to estimate $r_i(t)$, which is needed to compute $\theta_i(t)$. In a few cases, it was possible to obtain such estimates for earlier years.

In the case of the remaining two petroleum firms in Table 2.1, answers to these three questions were obtained for 1958 through correspondence with the research director. These firms were too far from Pittsburgh, New Haven, or Cambridge for interviews to be feasible.

One possible bias should be noted. Only about half of the firms that I contacted would give me the sort of information I need here, and the fact that certain firms would give me such data may indicate that they are more likely to conform to the model. If this is the case, there would obviously be much more unexplained variation if all firms were included in equation (10) in Chapter 2.

In addition, it is always possible that a firm might overstate the figure regarding its expectations of the profitability of R and D to rationalize large R and D expenditures carried out for other reasons. But this seems rather farfetched because the firms were not told what their estimates of $\bar{\rho}_i(t)$ and ρ^* would be used for. Moreover, it seems unlikely that they would go to this much trouble to deceive, when they could more easily claim that they could not answer.

Estimates of β_{i1}, β_{i2}, β_{i3}, β_{i4}, *and* $\tilde{R}_i(t)/S_i(t)$*:* The data on $R_i(t)$ and $R_i(t-1)$, based on the National Science Foundation definitions of R and D, come mainly from correspondence with the firms and from Langenhagen [72], but for a few firms they come from Horowitz [57]. Despite the instructions given the firms, it is possible that the data are not entirely comparable over time and among firms because somewhat different definitions of R and D were used. The data on $S_i(t)$ came from *Moody's*. For all firms, $R_i(t)$, $R_i(t-1)$, and $S_i(t)$ are measured in units of millions of dollars, and t is measured in years from 1945. In Table 2.4, we obviously are forced to omit firms where $[1 - \hat{\beta}_{i1} + \hat{\rho}_4] < 0$.

Although the least-squares estimates of β_{i1}, β_{i2}, β_{i3}, and β_{i4} are consistent, they are not unbiased because $R_i(t-1)$, a lagged endogenous variable, is used as an exogenous variable. See Hurwicz [58]. There seems to be no simple way to eliminate this bias, but fortunately it should be fairly small. According to Hurwicz's results, it would be about 10 percent; but his model differs from equation (15) in Chapter 2 since it includes no independent variables other than the lagged endogenous variable. Note too that, for a few firms, I could not get data regarding $R(t)$ for the entire period (1945–1958), and the first few years had to be omitted for this reason.

In addition, there are other biases. Since the expectation of the ratio of random variables does not equal the ratio of their expectations, the transformation from $\hat{\beta}_{i1}$, $\hat{\beta}_{i2}$, $\hat{\beta}_{i3}$, and $\hat{\beta}_{i4}$ to $\hat{\alpha}_{i1}$, $\hat{\alpha}_{i2}$, and $\hat{\alpha}_{i3}$ results in bias. Moreover, because $u_i(t)$ is omitted, the means in Table 2.4 would not necessarily equal the means of $\tilde{R}_i(t)/S_i(t)$, even if there were no errors in the estimates of α_{i1}, α_{i2}, and α_{i3}. However, under these circumstances, the means in Table 2.4 would be unbiased estimates. Similarly, the standard deviations of $\tilde{R}_i(t)/S_i(t)$ are underestimated because we ignore the variance of $u_i(t)$. Finally, the small number of glass and steel firms should be noted; the results pertain to only a few major firms.

Despite these problems, the general conclusions in section 3 of Chapter 2 are almost certainly correct; and although the particular numbers in Table 2.4 must be viewed with caution, there is evidence that they are reasonably accurate. When the direct estimates of $\tilde{R}_i(t)/S_i(t)$ in Table 2.1 are compared with those based on these assumptions, the results are quite close, the average estimates of $\tilde{R}_i(t)/S_i(t)$ differing by only 5 percent. Also, the estimates of β_{i1} agree quite well with the results in equation (11) in Chapter 2. According to the model, $\beta_{i1} = 1 + \nu_4 - \nu_5 - \nu_5\pi_i$; thus the average value of β_{i1} in each industry should equal $1 + \nu_4 - \nu_3 - \nu_5$, or 0.60 [according to equation (11)]. In fact, the average values of β_{i1} are 0.55 (chemicals), 0.50 (petroleum), 0.67 (drugs), 0.35 (steel), and 0.59 (glass). Taking account of sampling and other errors, the differences seem small.

Forecasts of R and D Expenditures: The root-mean-square error of the forecasts based on equation (15) in Chapter 2 is $1.07 million. Using the naïve forecast that expenditures in 1959 will equal those in 1958, the root-mean-square error is $2.10 million. Using the naïve forecast that expenditures in 1959 will differ from those in 1958 by the same amount that expenditures in 1958 differed from those in 1957, the root-mean-square error is $1.57 million. And using the naïve forecast that expenditures in 1959 will differ by the same percent from those in 1958 as 1958 expenditures differed from those in 1957, the root-mean-square error is $1.67 million. In all of these computations, firm C9 is omitted because the model in section 3 of Chapter 2 apparently does not apply in its case, the estimate of β_{i1} being significantly negative. It is included in the computations in section 2, but its omission there would make little difference to the results.

Our results seem to be better than forecasts based on businessmen's expectations. See Greenwald [45]. To obtain the forecasts where sales estimates contained errors, we distributed random errors of plus or minus 10 percent of sales.

For equation (16) in Chapter 2, the estimates of r were obtained as follows: The figure for petroleum is based on forecasts made by the president of Shell Oil in *The Wall Street Journal* (May 18, 1960), and by *Chemical Processing* in 1960. The figure for chemicals is an average of forecasts from *Chemical Week* (December 24, 1960) and *Chemical Processing.* The figure for drugs is from the president of Eli Lilly in *The Wall Street Journal* (May 18, 1960). The roughness of these forecasts of $S_i(t)$ need not be labored.

Estimates of n_i: Langenhagen [72] provides data on the number of significant inventions and the R and D expenditures of the chemical firms. The data on n_i come from his Table 2, Chapter 4. The inventions are weighted by the number of times they were included on the questionnaires he describes there. Three firms (cited by Langenhagen) which devoted a considerable amount of their R and D to fields other than chemicals had to be omitted for obvious reasons.

I looked up the names of those who were credited by various trade and technical sources with having invented each of the inventions in petroleum refining on Schmookler's list [145]. Then using biographical directories and other sources, I identified the firm that employed the inventor when the invention was made. (For most of the inventions during this period, I could find this information.) Next, I asked members of the Carnegie Institute of Technology engineering faculty to rank the inventions by their economic importance, and weighted them in proportion to their rank. Finally, I obtained similar data (described in Chapter 5) for petrochemical innovations and combined them with the data for process inventions.

For the steel firms, the data on n_i are described in Chapter 5 and Appendix B. Note that they, like the petrochemical data, pertain to innovations, not inventions. No data on the latter were available. Some of the data on R_i came from Ninian [123]; the rest are described in section 3 of Chapter 2.

Appendix B

Source of Tables 5.1, 5.2, and 5.3: Given the lists of innovations from the trade associations and trade journals, we could determine the identity of the innovator in about 90 percent of the cases in petroleum refining, 50 percent of the cases in steel, and practically all of the cases in coal. Although the data on the identity of the innovators are generally reliable, there are a few cases in each industry where the data—based on the recollection of suppliers, etc.—may possibly be wrong. To make sure that the lists were reasonably complete, they were checked with members of the Carnegie Institute of Technology engineering faculty and the Bureau of Mines. A few innovations were added to the lists (or dropped) on their recommendation. Next, we asked various trade journals and members of the Carnegie Institute of Technology engineering faculty whether, in their judgment, the results seemed to be biased in favor of large or small firms. No such bias could be detected, but this test is obviously rough. For further information regarding the sources of Table 5.1, see notes 7 and 8 of Chapter 6.

Source of Table 5.4: The daily crude capacity of each petroleum refiner (other than the top twenty companies) in 1927 was obtained from the *Petroleum Register*. For 1947 it was obtained from Bureau of Mines, *Petroleum Refineries, Including Cracking Plants in the United States*. The daily crude capacity (domestic and overseas) of the top twenty companies was obtained directly from the firms. The ingot capacity of each firm in 1926 and in 1945 was obtained from the *Directory of the American Iron and Steel Institute*.

For the coal industry, size distributions of firms in the "base states" are provided by H. Risser, *The Economics of the Coal Industry* (University of Kansas, 1958) for 1933 and 1953. We multiplied the number in each size class—under one million tons of production annually—by the ratio of the total production in the country to that in the "base states." A complete count of firms with over 1 million tons produced in 1933 and 1953 was obtained from the *Keystone Coal Buyer's Guide*. The production of the innovators was also obtained from this source. Risser's data exclude firms producing less than 1,000 tons annually.

In the steel industry, innovations introduced by firms without any ingot capacity or by firms engaged primarily in some other business had to be omitted. To include them on the basis of their ingot capacity would have been to misstate their true size. And no measure of size other than ingot capacity (or pig-iron capacity) is readily available for most of the firms in the industry. Such innovations are marked with a footnote in Table 5.1.

In the petroleum industry, a few innovations had to be omitted for much the same sort of reason. In the coal industry, innovations introduced by firms engaged primarily in some other business were not omitted. Such firms account for a large proportion of the industry's output, and the results would not have been altered much in any event if they had been omitted.

We follow the convention (adopted in most studies of industrial concentration) of using the largest four firms as a basis for concentration measures. In steel, the largest firms' share of the market is their share of the industry's ingot capacity. In petroleum, it is their share of the industry's daily crude capacity. In bituminous coal, it is their share of the industry's (tonnage) production.

Other measures—e.g., percent of value-added or percent of employment—might have been used instead. But census data for petroleum in 1935 and 1947 and for steel in 1947 indicate that the results would change only slightly. For 1919–1938, the largest petroleum firms' share of the innovations would have exceeded their share of value-added or employment, but the difference would have been somewhat smaller than in Table 5.4. For 1939–1958, the results in petroleum would have been about the same as in Table 5.4. For 1939–1958, the largest steel firms' share of the innovations would have been less than their share of the assets or employment, but larger than their share of value-added.

In 1919–1938, the difference between the largest firms' share of the innovations and their share of the market was almost always statistically significant in steel and petroleum. In 1939–1958, the share of the innovations introduced by the largest firms was closer to their share of the market than in 1919–1938, and often the differences may not have been statistically significant in steel and petroleum. In bituminous coal, there was a relatively small probability that the differences in Table 5.4 were due to chance in either period.

Source of Table 5.5: The data regarding M and $\bar{I}$ were obtained primarily from interviews with officials of engineering associations and firms, although some came from published sources. Estimates of M and I were obtained for as many of the innovations in Tables 5.1 and 5.2 as possible and the average values of M and I were used in Table 5.5. To some extent, the estimate of I for each innovation was also an average, since I would vary depending on a firm's existing plant. Of course, there was some variation among innovations in the value of M, the assumption in the text being only a convenient simplification.

Using the data described above, we determined $N(M)$. Since M was quoted in ingot capacity or crude capacity, we used the ratio of the largest firm's assets to its capacity in the mid-1950's to estimate M, $\bar{S}_M$, and $\bar{S}_4$ in terms of dollars (rather than capacity). This obviously is a very rough procedure. (For one thing, the capacities and assets sometimes pertained to somewhat different points in time. But if this is corrected, the results turn out to be almost exactly the same as those in equation (4) of Chapter 5.) The estimates of I are in (approximately) 1950 dollars. The weighted data regarding π are used.

The Estimated Effect of Breaking Up the Largest Firms: There are several obvious difficulties in the sort of analysis carried out in section 3 of Chapter 5.

(1) Although a firm's size influences the number of innovations it carries out, this is not the only factor. The preferences of its management with respect to risk, its profitability and rate of growth, and the size of its competitors may also be important. Thus, if the largest firms had been broken up, the behavior of their smaller successors might not have been like that of other smaller firms.

(2) If the largest firms had been broken up, the innovations that were introduced might have been of a different type. Only the largest firms may have been able to carry out some kinds of innovations. We assume that such innovations were no more important than those that their smaller successors would have introduced. There is no evidence in these industries that an innovation in Tables 5.1, 5.2, and 5.3 introduced by a larger firm tended to be any more—or less—important than one introduced by a smaller firm, but the data are very rough.

(3) If the largest firms had been broken up, changes might have occurred outside the industry. For example, if the largest firms had carried out a relatively large amount of research, there might have been some transfer of research activities to independent laboratories. The amount of inventive activity might not have been greatly affected, but the reorganization of the industry might have affected how many of the research results were applied—and how quickly.

(4) As the literature on cost and production functions clearly shows, there are many difficulties in interpreting least-squares relationships between a firm's size and other variables. Some of these difficulties are present here; e.g., there is an identification problem. For some reason, certain firms may be innovators, and as a consequence they may grow more rapidly than others. If so, they may eventually become relatively large and the largest may account for a disproportionately large share of the innovations—even though size per se brings no particular advantages. This hypothesis is obviously difficult to check.

The following points should be noted regarding the regressions. In steel, firms with less than 5,000 tons of ingot capacity were omitted in 1919–1938 and firms with less than 10,000 tons were omitted in 1939–1958. S_j was measured in units of 1,000 tons. In petroleum, firms with less than 500 barrels of capacity were omitted, and S_j was measured in units of 1,000 barrels. In coal, firms producing less than 100,000 tons annually were omitted, approximate size data were used for the 100,000- to 1,000,000-ton ranges, and S_j was measured in units of 1,000,000 tons.

Finally, note three other points. First, all of this pertains only to the existing ranges of firm size. There is no way to tell how firms bigger than the largest existing firm would have behaved. Second, since the size data are only approximate for coal firms producing less than 1,000,000 tons, we confined ourselves to the range above 1,000,000 tons when we looked for the maximum value of $N(S_j)/S_j$. Third, there are substantial sampling errors in the estimates of the a's and consequently in the estimates of the values of S_j where $N(S_j)/S_j$ is a maximum. Thus, on these grounds, too, the results should be treated with caution.

Appendix C

Iron and Steel Industry: For the continuous wide-strip mill, all firms having more than 140,000 tons of sheet capacity in 1926 were included in Chapter 7; for the by-product coke oven, all firms with over 200,000 tons of pig iron capacity in 1901 were included; and for continuous annealing of tin plate, the nine major producers of tin plate in 1935 were included. In the case of the strip mill and coke oven, a few of these firms merged or went out of business before installing them, and there was no choice but to exclude them.

The date when each firm first installed a continuous wide-strip mill was taken from the Association of Iron and Steel Engineers [7]. Similar data for the coke oven were obtained from various editions of the *Directory of Iron and Steel Works* of the American Iron and Steel Institute, and issues of the *Iron Trade Review* and *Iron Age*. The date when each firm installed continuous annealing lines was obtained from correspondence with the firms.

The size of the investment required and the durability of replaced equipment came from interviews. (The estimates of the pay-out periods were also checked there. Cf. note 20 of Chapter 7.) The interviews (each about 2 hours long) were with major officials of three steel firms, the president and research manager of a firm that builds strip mills and continuous annealing lines, officials of a firm that builds coke ovens, and representatives of a relevant engineering association and of a trade journal. The data on growth of output were annual industry growth rates and were for sheets during 1926–1937 (strip mill), pig iron during 1900–1925 (coke oven), and tin plate during 1939–1956 (continuous annealing). They were taken from the *Bituminous Coal Annual* of the Bituminous Coal Institute, Association of Iron and Steel Engineers [7], and *Annual Statistical Reports* of the American Iron and Steel Institute.

Railroad Industry: For centralized traffic control, all Class I line-haul roads with over 5 billion freight-ton miles in 1925, were included. For the diesel locomotive and car retarders, essentially the same firms were included. (The Norfolk and Western, a rather special case, was replaced by the New Haven and Lehigh Valley in the case of the diesel locomotive. Some important switching roads were substituted in the case of car retarders, an innovation in switching techniques.) An entire system is treated here as one firm.

The date when a firm first installed centralized traffic control was usually derived from a questionnaire filled out by the firm. For those that did not reply, estimates by K. Healy [52] were used. The date when each firm first installed diesel locomotives was determined from various editions of the

Interstate Commerce Commission's *Statistics of Railways*. The date when each firm first installed car retarders was taken from various issues of *Railway Age*.

For centralized traffic control and car retarders, some of the pay-out periods were estimates published in the *Signal Section, Proceedings of Association of American Railroads*, and the rest were obtained from questionnaires filled out by the firms. All estimates of the pay-out period for the diesel locomotive and the pay-out period required for investment were obtained from questionnaires. Information regarding the size of the investment required and the durability of old equipment was obtained primarily from interviews with eight officials of six railroads (ranging from president to chief engineer) and three officials of a signal manufacturing firm and a locomotive manufacturing firm. The data on growth of output were annual growth rates for total freight ton-miles during 1925–1941 (diesel locomotive) and 1925–1954 (centralized traffic control and car retarders). They were taken from the *Statistics of Railways*.

Bituminous Coal Industry: Practically all firms producing over 4 million tons of coal in 1956 (according to McGraw-Hill's *Keystone Coal Buyers Manual*) were included. A few firms that did strip mining predominantly were excluded in the case of the continuous mining machine, and a few had to be excluded in the case of the shuttle car and trackless mobile loader, because they would not provide the necessary data. The date when each firm first introduced these types of equipment was usually obtained from questionnaires filled out by the firms, but in the case of the continuous mining machine, data for two firms that did not reply were derived from the *Keystone Coal Buyers Manual*.

Data regarding the size of the investment required and the durability of old equipment were obtained from interviews with two vice-presidents of coal firms, several executives of firms manufacturing the equipment, employees of the Bureau of Mines, and representatives of an independent coal research organization. The data on growth of output were annual growth rates for bituminous coal production during 1934–1951 (trackless mobile loader), 1937–1951 (shuttle car), and 1947–1956 (continuous mining machine). They were taken primarily from the *Bituminous Coal Annual*.

Brewing Industry: We tried to include all breweries with more than $1 million in assets in 1934 (according to the *Thomas Register*), but several would not provide the necessary data and they could not be obtained elsewhere. The date when a firm first installed each type of equipment was usually taken from a questionnaire that it filled out, but in a few cases it was provided by manufacturers of the equipment or articles in the *Brewers' Journal*. The size of the necessary investment and the durability of equipment that was replaced were determined from interviews with a number of officials in two breweries and sales executives of two can companies. The data on growth of output were annual growth rates for beer production during 1935–1937 (tin containers) and 1950–1958 (pallet-loading machine and high-speed bottle filler). They were taken from the *Brewers' Almanac* and *Business Week* (June 20, 1959). The 1950–1958 figures refer only to half of these larger firms.

References

1. Adams, W., and J. Dirlam, "Big Steel, Invention, and Innovation," *Quarterly Journal of Economics* (May 1966).
2. American Mining Congress, *Coal Mine Modernization Yearbook* (annual).
3. Anderson, T., *An Introduction to Multivariate Statistical Analysis* (New York: Wiley, 1958).
4. Arrow, K., "The Economic Implications of Learning by Doing," *Review of Economic Studies* (June 1962).
5. Arrow, K., and W. Capron, "Dynamic Shortages and Price Rises: The Engineer-Scientist Case," *Quarterly Journal of Economics* (May 1959).
6. Arrow, K., H. Chenery, B. Minhas, and R. Solow, "Capital-Labor Substitution and Economic Efficiency," *Review of Economics and Statistics* (Aug. 1961).
7. Association of Iron and Steel Engineers, *The Modern Strip Mill* (Pittsburgh, 1941).
8. Bailey, N., "Some Problems in the Statistical Analysis of Epidemic Data," *Journal of the Royal Statistical Society*, B. 17 (1955).
9. Bain, J., *The Economics of the Pacific Coast Petroleum Industry* (Berkeley: University of California, 1944), Vol. 3.
10. ———, *Pricing, Distribution, and Employment* (New York: Henry Holt, 1953).
11. Baker, N., and W. Pound, "R and D Project Selection: Where We Stand," presented at the joint TIMS-ORSA Meeting, Minneapolis, Oct. 1964.
12. Beal, G., and J. Bohlen, *The Diffusion Process*, Special Report No. 18, Agricultural Extension Service (Ames: Iowa State College, 1957).
13. Berkson, J., "A Statistically Precise and Relatively Simple Method of Estimating the Bio-Assay with Quantal Response, Based on the Logistic Function," *Journal of the American Statistical Association* (Sept. 1953).
14. Blank, D., and G. Stigler, *The Demand and Supply of Scientific Personnel* (New York: National Bureau of Economic Research, 1957).
15. Bohlen, J., and G. Beal, *How Farm People Accept New Ideas*, Special Report No. 15, Agricultural Extension Service (Ames: Iowa State College, 1955).
16. Brandenburg, R., "Research and Development Project Selection" (Ph.D. thesis, Cornell, 1964).
17. Brown, W., "Innovation in the Machine Tool Industry," *Quarterly Journal of Economics* (Aug. 1957).

18. Brozen, Y., "Invention, Innovation, and Imitation," *American Economic Review* (May 1951).
19. ———, "Trends in Industrial Research and Development," *Journal of Business* (July 1960).
20. ———, "The Future of Industrial Research," *Journal of Business* (Oct. 1961).
21. Camp, J., and C. Francis, *The Making, Shaping, and Treating of Steel* (U.S. Steel Co., 1940).
22. Carter, C., and B. Williams, *Industry and Technical Progress* (New York: Oxford University Press, 1957).
23. Cattell, J., "The Influence of the Intensity of the Stimulus on the Length of the Reaction Time," reprinted in Dennis, *Readings in the History of Psychology* (New York: Appleton-Century-Crofts, 1948).
24. Chenery, H., "Overcapacity and the Acceleration Principle," *Econometrica* (Jan. 1952).
25. Coleman, J., E. Katz, and H. Menzel, "The Diffusion of an Innovation Among Physicians," *Sociometry* (Dec. 1957).
26. Conference on Price Research, *Cost Behavior and Public Policy* (New York: National Bureau of Economic Research, 1943).
27. Creamer, D., "Postwar Trends in the Relation of Capital to Output in Manufactures," *American Economic Review* (May 1958).
28. Creamer, D., S. Dobrovolsky, and I. Borenstein, *Capital in Manufacturing and Mining* (Princeton, 1960).
29. Dean, B., and S. Sengupta, "Research Budgeting and Project Selection," *IRE Transactions on Engineering Management* (Dec. 1962).
30. Domar, E., "Investment, Losses, and Monopoly," *Income, Employment, and Public Policy* (New York: Norton, 1948).
31. ———, "On Total Productivity and All That," *Journal of Political Economy* (Dec. 1962).
32. Eisner, R., "A Distributed Lag Investment Function," *Econometrica* (Jan. 1960).
33. Enos, J., "The History of Cracking in the Petroleum Refining Industry" (Ph.D. thesis, M.I.T., 1958).
34. ———, "Invention and Innovation in the Petroleum Refining Industry," *The Rate and Direction of Inventive Activity* (Princeton, 1962).
35. Ewell, R., "Role of Research in Economic Growth," *Chemical Engineering News* (July 1955).
36. Fellner, W., "The Influence of Market Structure on Technological Progress," *A.E.A. Readings in Industrial Organization and Public Policy* (Homewood, Ill.: Irwin, 1958).
37. Foell, C., and M. Thompson, *The Diesel Electric Locomotive* (Diesel Publications, Incorporated, 1946).
38. Freeman, R. J., "A Stochastic Model for Determining the Size and Allocation of the Research Budget," *IRE Transactions on Engineering Management* (Mar. 1960).

39. Gainsbrough, M., "Allocation of Resources to Research and Development," *Conference on Research and Development* (National Science Foundation, 1958).
40. Galbraith, J., *American Capitalism* (Boston: Houghton Mifflin, 1952).
41. Gilfillan, S., *Inventing the Ship* (Chicago: Follett, 1935).
42. Goodwin, R., "Secular and Cyclical Aspects of the Multiplier and the Accelerator," *Income, Employment, and Public Policy* (New York: Norton, 1948).
43. Gordon, M., "The Payoff Period and the Rate of Profit," *Journal of Business* (Oct. 1955).
44. Graue, E., "Inventions and Production," *Review of Economics and Statistics* (Nov. 1953).
45. Greenwald, D., "The Annual McGraw-Hill Research and Development Survey," *Methodology of Statistics on Research and Development* (National Science Foundation, 1959).
46. Griliches, Z., "Hybrid Corn: An Exploration in the Economics of Technological Change," *Econometrica* (Oct. 1957).
47. ———, "Research Costs and Social Returns: Hybrid Corn and Related Innovations," *Journal of Political Economy* (Oct. 1958).
48. ———, "Research Expenditures, Education, and the Aggregate Agricultural Production Function," *American Economic Review* (Dec. 1964).
49. ———, "Comment," *American Economic Review* (May 1965).
50. Hamberg, D., *Testimony on Employment, Growth, and Price Levels Before the Joint Economic Committee of Congress* (1959).
51. ———, "Invention in the Industrial Research Laboratory," *Journal of Political Economy* (Apr. 1963).
52. Healy, K., "Regularization of Capital Investment in Railroads," *Regularization of Business Investment* (New York: National Bureau of Economic Research, 1954).
53. Hennipman, P., "Monopoly: Impediment or Stimulus to Economic Progress," *Monopoly and Competition and Their Regulation* (New York: Macmillan, 1954).
54. Hildebrand, P., and E. Partenheimer, "Socioeconomic Characteristics of Innovators," *Journal of Farm Economics* (May 1958).
55. Hitch, C., and R. McKean, *The Economics of Defense in the Nuclear Age* (Cambridge: Harvard, 1960).
56. Hodges, J., "A Report on the Calculation of Capital Coefficients for the Petroleum Industry," *Problems of Capital Formation* (Princeton, 1957).
57. Horowitz, I., "Regression Models for Company Expenditures on and Returns from Research and Development," *IRE Transactions on Engineering Management* (1960).
58. Hurwicz, L., "Least-Squares Bias in Time Series," *Statistical Inference in Dynamic Economic Models*, ed. by T. Koopmans (New York: Wiley, 1950).

59. Interstate Commerce Commission, *Statistics of Railways* (annual).
60. Jerome, H., *Mechanization in Industry* (New York: National Bureau of Economic Research, 1934).
61. Jewkes, J., D. Sawers, and R. Stillerman, *The Sources of Invention* (New York: St. Martin's Press, 1958).
62. Kaplan, A., *Big Enterprise in a Competitive System* (Washington, D.C.: Brookings Institution, 1954).
63. Keezer, D., D. Greenwald, and R. Ulin, "The Outlook for Expenditures on Research and Development During the Next Decade," *American Economic Review* (May 1960).
64. Keirstead, B., *The Theory of Economic Change* (New York: Macmillan, 1948).
65. Kendall, M. G., *The Advanced Theory of Statistics* (New York: Hafner, 1951).
66. Kendrick, J., *Productivity Trends in the United States* (New York: National Bureau of Economic Research, 1961).
67. Kisselgoff, A., and F. Modigliani, "Private Investment in the Electric Power Industry and the Acceleration Principle," *Review of Economics and Statistics* (Nov. 1957).
68. Klein, B., "The Decision Making Problem in Development," *The Rate and Direction of Inventive Activity* (New York: National Bureau of Economic Research, 1962).
69. Klein, L., "Studies in Investment Behavior," *Conference on Business Cycles* (New York: National Bureau of Economic Research, 1951).
70. Kurz, M., "Research and Development, Technical Change, and the Competitive Mechanism," Institute of Mathematical Studies in the Social Sciences (Stanford, 1962).
71. Kuznets, S., "Inventive Activity: Problems of Definition and Measurement," *The Rate and Direction of Inventive Activity* (Princeton, 1962).
72. Langenhagen, C., "An Evaluation of Research and Development in the Chemical Industry" (unpublished M.S. thesis, M.I.T., 1958).
73. Lilienthal, D., *Big Business: A New Era* (New York: Harpers, 1953).
74. Lynn, F., "An Investigation of the Rate of Development and Diffusion of Technology in Our Modern Industrial Society," *Report to the President of the National Commission on Technology, Automation, and Economic Progress* (Washington, 1966).
75. McGraw-Hill, *Business Plans for Expenditures on Plant and Equipment* (annual).
76. Machlup, F., "The Supply of Inventors and Inventions," *The Rate and Direction of Inventive Activity* (Princeton, 1962).
77. Mack, R., *The Flow of Business Funds and Consumer Purchasing Power* (New York: Columbia, 1941).
78. MacLaurin, W., "Technological Progress in Some American Industries," *Quarterly Journal of Economics* (Feb. 1953).

79. Malcolm, D., J. Rosebloom, C. Clark, and W. Fagar, "Application of a Technique for R and D Project Evaluation," *Operations Research* (Sept. 1959).
80. Mandelbrot, B., "New Methods in Statistical Economics," *Journal of Political Economy* (Oct. 1963).
81. Mansfield, E., "Technical Change and the Rate of Imitation," *Econometrica* (Oct. 1961).
82. ———, "Comment," *The Rate and Direction of Inventive Activity* (Princeton, 1962).
83. ———, "Entry, Gibrat's Law, Innovation, and the Growth of Firms," *American Economic Review* (Dec. 1962).
84. ———, "Comment," *American Economic Review* (May 1963).
85. ———, "Intrafirm Rates of Diffusion of an Innovation," *Review of Economics and Statistics* (Nov. 1963).
86. ———, "Size of Firm, Market Structure, and Innovation," *Journal of Political Economy* (Dec. 1963).
87. ———, "The Speed of Response of Firms to New Techniques," *Quarterly Journal of Economics* (May 1963).
88. ———, "Industrial Research and Development Expenditures: Determinants, Prospects, and Relation to Size of Firm and Inventive Output," *Journal of Political Economy* (Aug. 1964).
89. ———, *Monopoly Power and Economic Performance* (New York: Norton, 1964).
90. ———, "The Economics of Research and Development: A Survey of Issues, Findings, and Needed Future Research," *Patents and Progress* (Homewood, Ill.: Irwin, 1965).
91. ———, "Economics, Public Policy and the Patent System," *Journal of the Patent Office Society* (May 1965).
92. ———, "Innovation and Technical Change in the Railroad Industry," *Transportation Economics* (National Bureau of Economic Research, 1965).
93. ———, "Rates of Return from Industrial Research and Development," *American Economic Review* (May 1965).
94. ———, "Technical Change and the Management of Research and Development," *Technological Change and Economic Growth* (Ann Arbor: University of Michigan Press, 1965).
95. ———, "National Science Policy," *American Economic Review* (May 1966).
96. ———, "Technological Change: Measurement, Determinants, and Diffusion," *Report to the President of the National Commission on Technology, Automation, and Economic Progress* (1966).
97. ———, *The Economics of Technological Change* (New York: Norton, 1968).
98. Mansfield, E., and R. Brandenburg, "The Allocation, Characteristics,

and Success of the Firm's R and D Portfolio: A Case Study," *Journal of Business* (Oct. 1966).
99. Mansfield, E., and C. Hensley, "The Logistic Process: Epidemic Curve and Applications," *Journal of the Royal Statistical Society*, B, 22 (1960).
100. Mansfield, E., and H. Wein, "A Model for the Location of a Railroad Classification Yard," *Management Science* (Apr. 1958).
101. March, J., and H. Simon, *Organizations* (New York: Wiley, 1958).
102. Marcson, S., *The Scientist in American Industry* (New York: Harpers, 1960).
103. Marglin, S., *Approaches to Dynamic Investment Planning* (Amsterdam: North-Holland Publishing Co., 1963).
104. Markowitz, H., *Portfolio Selection* (New York: Wiley, 1959).
105. Marschak, T., "Strategy and Organization in a System Development Project," *The Rate and Direction of Inventive Activity* (Princeton, 1962).
106. Marshall, A., and W. Meckling, "Predictability of Costs, Time and Success of Development," *The Rate and Direction of Inventive Activity* (New York: National Bureau of Economic Research, 1962).
107. Mason, E., "Schumpeter on Monopoly and the Large Firm," *Review of Economics and Statistics* (May 1951).
108. Massell, B., "A Disaggregated View of Technical Change," *Journal of Political Economy* (Dec. 1961).
109. Minasian, J., "Technical Change and Production Functions," unpublished paper presented at the fall 1961 meetings of the Econometric Society.
110. ———, "The Economics of Research and Development," *The Rate and Direction of Inventive Activity* (Princeton, 1962).
111. Modigliani, F., "Comment," *Problems in Capital Formation* (Princeton, 1957).
112. Modigliani, F., and H. Weingartner, "Forecasting Uses of Anticipatory Data on Investment and Sales," *Quarterly Journal of Economics* (Feb. 1958).
113. Moore, G., "Measuring Recessions," *Journal of the American Statistical Association* (June 1958).
114. Mottley, C., and R. Newton, "The Selection of Projects for Industrial Research," *Operations Research* (Nov. 1959).
115. Mueller, W., "A Case Study of Product Discovery and Innovation Costs," *Southern Economic Journal* (July 1957).
116. National Commission on Technology, Automation, and Economic Progress, *Technology and the American Economy* (Washington, 1966).
117. Nelson, R., "The Economics of Invention: A Survey of the Literature," *Journal of Business* (Apr. 1959).
118. ———, "The Simple Economics of Basic Scientific Research," *Journal of Political Economy* (June 1959).
119. ———, "Uncertainty, Learning, and the Economics of Parallel Re-

search and Development Efforts," *Review of Economics and Statistics* (Nov. 1961).

120. ———, "The Link Between Science and Invention: The Case of the Transistor," *The Rate and Direction of Inventive Activity* (Princeton, 1962).
121. ———, "The Allocation of Research and Development Resources," *The Economics of Research and Development* (Columbus: Ohio State University Press, 1965).
122. Nelson, R., M. Peck, and E. Kalachek, *Technology, Economic Growth, and Public Policy* (Washington, D.C.: The Brookings Institution, 1966).
123. Ninian, A., "The Role of Research in the American Steel Industry" (unpublished M.S. thesis, M.I.T., 1959).
124. Nutter, G., "Monopoly, Bigness, and Progress," *Journal of Political Economy* (Dec. 1956).
125. Organization for European Economic Cooperation, *The Organization of Applied Research in Europe, the United States, and Canada* (OEEC, 1954).
126. Peck, M., "Inventions in the Postwar American Aluminum Industry," *The Rate and Direction of Inventive Activity* (Princeton, 1962).
127. Perles, B., "Innovation in the Machine Tool Industry: Comment," *Quarterly Journal of Economics* (Aug. 1963).
128. Phelps, E., "The New View of Investment: A Neoclassical Analysis," *Quarterly Journal of Economics* (Nov. 1962).
129. ———, "Substitution, Fixed Proportions, Growth, and Distribution," *International Economic Review* (Sept. 1963).
130. Pound, W. H., "Research Project Selection: Testing a Model in the Field," *IRE Transactions on Engineering Management* (Mar. 1964).
131. Quinn, J., "Long-Range Planning of Industrial Research," *Harvard Business Review* (July 1961).
132. *Reviews of Data on Research and Development*, National Science Foundation, 1956 to present.
133. Robinson, J., *The Rate of Interest* (New York: Macmillan, 1952).
134. ———, *The Accumulation of Capital* (Homewood, Ill.: Irwin, 1956).
135. Rubenstein, A., "Setting Criteria for R and D," *Harvard Business Review* (Jan. 1957).
136. ———, "Rate of Organizational Change, Corporate Decentralization, and the Constraints on Research and Development in the Firm," paper presented at the Institute of Management Science (June 1959).
137. Rubenstein, A., and R. Hannenberg, "Idea Flow and Project Selection in Several Industrial Research and Development Laboratories," *The Economics of Research and Development* (Ohio State, 1965).
138. Salter, W., *Productivity and Technical Change* (Cambridge: Cambridge University, 1960).
139. Sanow, K., "Development of Statistics Relating to Research and De-

velopment Activities in Private Industry," *Methodology of Statistics on Research and Development* (National Science Foundation, 1959).
140. Scherer, F., "Comment," *The Rate and Direction of Inventive Activity* (National Bureau of Economic Research, 1962).
141. ———, "Size of Firm, Oligopoly, and Research: A Comment," *Canadian Journal of Economics and Political Science* (May 1965).
142. ———, "Firm Size, Market Structure, Opportunity, and the Output of Patented Inventions," *American Economic Review* (Dec. 1965).
143. Schmookler, J., "Inventors Past and Present," *Review of Economics and Statistics* (Aug. 1957).
144. ———, "Bigness, Fewness, and Research," *Journal of Political Economy* (Dec. 1959).
145. ———, "Changes in Industry and in the State of Knowledge as Determinants of Inventive Activity," *The Rate and Direction of Inventive Activity* (Princeton, 1962).
146. Schmookler, J., and O. Brownlee, "Determinants of Inventive Activity," *American Economic Review* (May 1962).
147. ———, "Invention, Innovation, and Business Cycles," *Some Elements Shaping Investment Decisions* (Joint Economic Committee of Congress, 1962).
148. ———, *Invention and Economic Growth* (Cambridge: Harvard, 1966).
149. Schumpeter, J., *Business Cycles* (New York: McGraw-Hill, 1939).
150. ———, *Capitalism, Socialism, and Democracy* (New York: Harper and Row, 1942).
151. Scitovsky, T., "Economic Theory and the Measurement of Concentration," *Business Concentration and Price Policy* (Princeton, 1955).
152. ———, "Economics of Scale and European Integration," *American Economic Review* (Mar. 1956).
153. Slutsky, E., "The Summation of Random Causes as the Source of Cyclic Processes," *Econometrica* (Apr. 1937).
154. Solow, R., "Technical Change and the Aggregate Production Function," *Review of Economics and Statistics* (Aug. 1957).
155. ———, "Investment and Technical Progress," *Mathematical Methods in the Social Sciences* (Stanford, 1959).
156. ———, "Capital, Labor, and Income in Manufacturing," *The Behavior of Income Shares* (Princeton, 1964).
157. Stelzer, I., "Technical Progress and Market Structure," *Southern Economic Journal* (July 1956).
158. Stigler, G., "Industrial Organization and Economic Progress," *The State of the Social Sciences* (Chicago: University of Chicago Press, 1956).
159. ———, *Testimony Before Subcommittee on Study of Monopoly Power*, Judiciary Committee, House of Representatives, 1950.
160. Steever, D., *Capacity Utilization and Business Investment* (University of Illinois, 1960).

161. Stocking, G., *Testimony Before Subcommittee on Study of Monopoly Power*, Judiciary Committee, House of Representatives, 1950.
162. Sutherland, A., "The Diffusion of an Innovation in Cotton Spinning," *The Journal of Industrial Economics* (Mar. 1959).
163. Swalm, F., "On Calculating the Rate of Return of an Investment," *Journal of Industrial Engineering* (Mar. 1958).
164. Terborgh, G., *Dynamic Equipment Policy* (New York: McGraw-Hill, 1949).
165. Terleckyj, N., "Sources of Productivity Advance" (Ph.D. thesis, Columbia, 1960).
166. Union Switch and Signal Co., *Centralized Traffic Control* (Swissvale, 1931).
167. Villard, H., "Competition, Oligopoly, and Research," *Journal of Political Economy* (Dec. 1958).
168. Worley, J., "Industrial Research and the New Competition," *Journal of Political Economy* (Apr. 1961).
169. Yance, J., "Technological Change as a Learning Process: The Dieselization of the Railroads" (unpublished, 1957).

Index